The Boundary Element Method

To our wives:
Shayla Ali and Lalitha Rajakumar

and our children:
Aleef & Teeasha Ali, and Vinod & Anita Rajakumar

The Boundary Element Method

Applications in Sound and Vibration

Ashraf Ali
Engineering Solution and Support (ESAS), Bellevue, Washington
(Formerly with Ansys, Inc., Canonsburg, Pennsylvania)

Charles Rajakumar
Ansys, Inc., Canonsburg, Pennsylvania

CRC Press
Taylor & Francis Group
Boca Raton London New York

CRC Press is an imprint of the
Taylor & Francis Group, an **informa** business
A BALKEMA BOOK

Library of Congress Cataloging-in-Publication Data

A Catalogue record for this book is available from the Library of Congress

Cover design: Miranda Bourgonjen
Typesetting: Charon Tec Pvt. Ltd, Chennai, India

CRC Press
Taylor & Francis Group
6000 Broken Sound Parkway NW, Suite 300
Boca Raton, FL 33487-2742

First issued in hardback 2020

ISBN-13: 978-90-5809-657-9 (hbk)

Contents

Preface

The boundary element method is a powerful discretization tool in computational mechanics. However, the eigenvalue analysis procedures within the boundary element discretization process are still in a developing stage. To our knowledge, this is the first-ever book dedicated entirely to the subject of boundary element eigenvalue formulations. All the techniques of boundary element eigenvalue analysis currently available in the literature are reviewed and presented in the book. For each technique, a detailed theoretical formulation is presented, followed by numerical illustrations. The advantages and disadvantages of each method in terms of computational efficiencies, generalities, and formulation difficulties are also presented.

The book includes detailed formulations on linear and quadratic eigensolvers for unsymmetric matrices since boundary element matrices are naturally unsymmetric. The book also sheds light on the ongoing debate on the choice of technique, the relative merits of eigenanalyses based on the boundary element and finite element methods, the unresolved issues that require immediate attention and the future direction of research in this area.

The mode–frequency analyses of vibrating structures and the computation of resonant frequencies of acoustical cavities are now routinely performed in the industry. The eigenanalysis based on the boundary element method holds promise of becoming a user-friendly and popular procedure with practicing engineers simply because here it can avoid the tedious and time-consuming process of creating an adequate mesh for their models. Some applications include:

Elasticity Area:

1. Machines (automobiles, aircraft, etc.);
2. Machine parts such as connectors, shafts, gears, fasteners such as screws, pins, etc;
3. Other equipment that is subjected to vibrations during normal operation;
4. Structures (bridges, buildings, etc.).

Acoustics Area:

1. Acoustic enclosures such as auditoriums, theaters, passenger-car and train cabins, etc.;
2. Hi-Fi sound equipment such as loudspeakers;
3. Fluid-filled structures such as oil tankers;
4. Noise control of structures such as automobile mufflers; aircraft fuselages, rooms housing vibrating machines, etc.

Furthermore, eigenvalue analysis forms the basis for subsequent mode-based dynamic analyses, such as mode superposition transient analysis, spectrum analysis, and random vibration analysis.

Before the advent of practical numerical methods like the finite element method, engineers conducted experiments on prototypes for determining natural frequencies. Starting in the 1970s, computer programs based on the finite element method were available. Although the finite element method is a versatile computational technique, it requires a much longer data-preparation time than the boundary element method. Specifically, engineers are forced to spend a significant amount of time in generating an adequate mesh for the model problem. Despite the introduction of a few automatic mesh-generation algorithms in the commercial finite element programs, engineers still continue to struggle in creating meshes of acceptable quality.

The boundary element method, on the other hand, is a boundary technique where only the boundary of the domain is required to be meshed, thereby simplifying data preparation efforts significantly. Among other benefits, the overall physical time spent by engineers to perform the analysis is reduced significantly, and the analysis process becomes more user-friendly.

The book presents the eigenvalue analysis techniques that use the boundary element method. The boundary element method does not easily lend itself to eigenvalue formulations, especially algebraic eigenvalue formulations. Consequently, publications on boundary element algebraic eigenvalue formulations did not come out until the mid-1980s, although the boundary element method has been around since the late 1960s. However, non-algebraic boundary element eigenvalue analysis, which is not a practical analysis technique, appeared in the literature of the mid-1970s. For historical reasons, the book presents some materials related to the non-algebraic boundary element eigenvalue analysis techniques.

Three general purpose boundary element computer programs (GPBEST, BEASY, and SYSNOISE) offer boundary element eigenvalue analysis capabilities. This book will hopefully satisfy the needs of engineers to acquire a detailed knowledge on the subject. The capabilities of the commercial programs, such as those mentioned, may be enhanced through the implementation of some of the different methods of performing boundary element eigenvalue analysis presented in the book. This book should also encourage the development of new and more powerful computer programs on boundary element eigenvalue analysis.

The book can be used in the graduate classes on "computational mechanics" and "boundary element methods". The researchers in universities and industries, practicing engineers, mathematicians, computer scientists, physicists, chemists and chemical engineers, and researchers in bio-medical fields can also use it as a reference.

Ashraf Ali
Seattle, Washington

Charles Rajakumar
Pittsburgh, Pennsylvania

Acknowledgements

We are indebted to Ansys, Inc., Canonsburg, Pennsylvania for getting us interested in the subject of boundary element method. We are also grateful to Sidney Solomon of The Solomon Press of New York in giving us encouragement and valuable suggestions in the preparation and completion of the book.

Abbreviations

1-D	One-dimensional, One dimension
2-D	Two-dimensional, Two dimensions
3-D	Three-dimensional, Three dimensions
BE	Boundary Element
BEM	Boundary Element Method
BIEM	Boundary Integral Equation Method
CPU	Central Processing Unit
DRM	Dual Reciprocity Method
DSM	Determinant Search Method
FE	Finite Element
FEM	Finite Element Method
FFM	Fictitious Function Method
GSF	Global Shape Function
ICM	Internal Cell Method
MRM	Multiple Reciprocity Method
NACA	National Advisory Committee for Aeronautics
PIM	Particular Integral Method
PSF	Polynomial Shape Function
SEM	Series Expansion Method

Chapter 1

Introduction

1.1. Why the boundary element method?

In the last thirty to thirty-five years the Boundary Element Method (BEM) has emerged as one of the most powerful computational tools for solving a wide variety of problems in science and engineering. While the Finite Element Method (FEM) is known to be versatile, the BEM brings with it the extraordinary feature of being simple in geometric data preparation. This particular feature of BEM derives from the fact that the discretization of the problem domain is confined to the boundary alone, i.e., the unknowns to be solved for are only on the boundary. The solution inside the domain can be computed as a post-processing step after the unknowns on the boundary points have been solved for.

In the FEM, the entire domain must be discretized in order to set up the algebraic equations and get solutions. It not only increases the number of equations that must be solved, but also burdens the user with generating an adequate mesh on the surface as well as in the interior of the domain. Despite the advent of a number of algorithms of automatic mesh generation to be used with the FEM, the users of the FEM today are still forced to allocate more than half of their time in creating suitable meshes for their problems.

Since the BEM reduces the problem dimension by one, two-dimensional (2-D) problems can be solved in one dimension and three-dimensional (3-D) problems can be posed in two dimensions. Therefore, only a line mesh around the boundary of the domain is needed in two dimensions and a surface mesh for 3-D geometries. It leads to dramatic reductions in mesh generation efforts, resulting in significant savings in processing time put in by an engineer toward solving the problem at hand. This particular property of the BEM makes it an attractive numerical analysis tool.

The BEM is an integral-type of numerical analysis procedure in which the integration of the governing differential equations is performed before the numerical analysis has been carried out. The FEM, on the other hand, is a differential-type numerical analysis technique because the numerical analysis part is performed first followed by the integration of the governing differential equation. The FEM may also be designated as a local technique. Here the entire problem domain is divided into "finite elements," which form the building blocks for reconstructing the whole domain. The numerical analysis is performed on the individual elements. The finite elements are then assembled for the entire domain, which is equivalent to the integration of the governing differential equation. The compatibility between adjacent elements is ensured during the process of assembly of the element matrices, and the equilibrium of the individual element ensures the overall equilibrium of the whole domain after assembly.

The BEM is a global numerical analysis procedure. The solution of the problem is found by superposing singular solutions distributed over the entire boundary of the problem. The singular source located at one point of the boundary exerts influence on each and every point on the boundary of the problem. When this influence of a single source for a discretized boundary is summed over all the boundary segments/elements, it fills the entire row of the final algebraic matrix equation. Therefore, a separate assembly procedure is not called for. The equilibrium is globally satisfied at once for the whole domain.

The BEM is more efficient than the FEM for several classes of problems, viz., infinite-domain problems such as those in acoustics, electrostatics, and electromagnetics, and problems with stiff gradients such as those in fracture mechanics. In some cases, a combined BEM–FEM procedure, in which the strengths of both methods can be exploited, is found to be optimal. The FEM is known to handle inhomogeneities and nonlinearities in the domain more efficiently. Therefore, the part of the domain that contains inhomogeneities and/or nonlinearities can be modeled using the finite elements, whereas the part that is homogeneous and/or extends to infinity may be modeled using boundary elements.

1.2. Typical applications of the boundary element method

Consider the return-and-go conductor problem, also known as the magnetic dipole problem, in which two conductors in free space carry current in opposite directions to infinity. The problem is to compute the magnetic flux density distribution both inside and outside of the conductors. Figure 1.1 shows four different ways of solving it. In the first method, the problem is solved using the BEM alone. Only the boundaries of the conductors are required to be discretized in order to model both the interiors of the conductors and the infinite-extent external domain. In the second method, the interior of one of the conductors is meshed using finite elements, whereas the interior of the other conductor as well as the infinite-extent exterior domain are modeled with the help of boundary elements. In the third method, the interiors of both conductors are modeled using finite elements, while the infinite-extent external domain is modeled using boundary elements. In the fourth method, the problem is solved using finite elements for both conductors and a portion of the external domain and boundary elements beyond.

In the first three cases, the zero-potential and zero-flux boundary conditions at infinity are implicitly satisfied by the boundary elements, although the discretization is confined to the surface of the conductors. Since the BEM is a global technique, the conductors that are physically disconnected at the two-dimensional (2-D) plane are easily modeled without requiring the domain between the conductors to be discretized. Also, in the first case, both the interior and the exterior domains are modeled using just one discretization at the conductor boundaries. In other words, in the BEMs, the boundary discretization used to model the interior domain can be used to model the external domain just by flipping the outward normal. Note that because of symmetry and anti-symmetry, the return-and-go conductor problem is, in practice, solved using only a quarter of the domain.

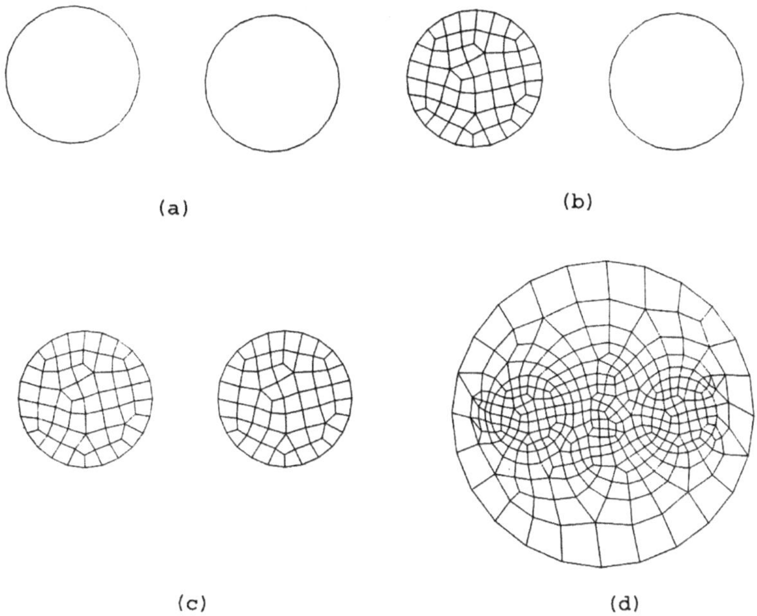

Figure 1.1. The return-and-go conductor problem. (a) Both conductors and exterior domain are modeled using BE alone; (b) interior of one conductor and exterior domain are modeled using BE while the interior of the other conductor is modeled using FE; (c) interiors of both conductors are modeled using FE, while the exterior domain is modeled using BE; (d) interiors of both conductors and a portion of the exterior domain are modeled using FE while the exterior domain is modeled using BE.

Over the years, BEMs have been applied to many branches of engineering science, such as: heat conduction, elastostatics, elastodynamics, elastoplasticity, viscoplasticity, acoustics, fracture mechanics, fluid flow, fluid–structure interaction problems, and electromagnetics.

However, the eigenvalue analysis formulations in the context of BEMs did not appear before the late 1970s. This is because the BEM does not easily lend itself to an algebraic eigenvalue formulation. The evolutionary history of different types of eigenvalue formulations with the BEMs will be presented later in this chapter. Before that, a brief chronological history of the emergence of the BEM itself is presented below.

1.3. Emergence of the boundary element method

As mentioned earlier, the BEM is an integral equation technique. The study of the integral equations started many decades before the boundary integral equation method (BIEM) emerged as a practical numerical analysis technique. In 1903, Fredholm [1] published his work on the application of integral equations to the formulation of boundary-value problems in potential theory. Early works on the integral equations were restricted to the study of existence and uniqueness of solutions to the problems encountered in mathematical physics. Trefftz [2] and Prager [3] developed methods to solve integral equations in potential fluid flow problems. These methods are actually suited for computers and were not of much use in those days. However, they may be called the precursors of modern BIEMs.

Kellog [4] applied integral equations to the solution of problems governed by Laplace's equation. Boundary integral equations were set up using integral transformation theorems to represent a harmonic function by superposing a single-layer and a double-layer potential. After specializing the equation on the boundary of the domain, the Fredholm integral equation of the second kind, relating the harmonic function and its derivative as unknowns on the boundary, could be established. Its counterpart in the theory of elasticity is the Somigliana identity [5], which relates the boundary displacement and boundary traction through an integral identity. The Russian author Muskhelishvili [6, 7] applied integral equations to the theory of elasticity. He used the complex variable method, and as such the application was restricted to 2-D. In 1957, another Russian author, Mikhlin [6], studied the properties of integral equations.

Smith and Pierce [7] used the "indirect" BIEM to study potential fluid flow problem. The indirect BIEM uses non-physical "source densities" as the unknowns on the boundary to be solved for. The physical variables anywhere in the domain are solved afterwards in terms of the source densities. The indirect methods were traditionally used in the solution of general potential and fluid flow problems. Friedman and Shaw [10, 11] and Shaw [12] in 1962, and Banaugh and Goldsmith [13] in 1963 applied the "direct" boundary integral method in acoustics to study the acoustic scattering problem. Hess [14] and Hess and Smith [15] calculated potential flow around bodies utilizing indirect boundary integral equations.

Jawson [16] and Symm [17] published their two-part paper on integral equation methods in potential theory. In these papers, they presented a numerical method in which they divided the problem boundary into small segments and assumed the unknown quantities to remain constant over the segments (so-called "constant" boundary elements). The integrals over the segments were computed using Simpson's rule. The singular integrals were treated separately. This led to a system of algebraic equations. Jawson and Symm solved simple 2-D potential problems using this procedure. Jawson and Ponter [18] applied this technique to solve torsion problems. Massonnet [19] also solved torsion problems numerically using the integral equation technique. In 1965, Kupradze [20] formulated *vector* integral equations, similar to those of Fredholm in potential theory, for applications in the theory of elasticity. Mikhlin [21, 22] proposed approximate solution techniques for solving integral equations and also presented multidimensional or vector integral equations.

In 1967, Jawson et al. [23], Rim and Henry [24], and Rizzo [25] applied the integral equation method to solve problems in elasticity. Oliveira [26] also performed plane stress analysis in elasticity with the help of the integral equation technique. Cruise and Rizzo [27] and Cruise [28] presented a boundary integral equation formulation for numerically solving transient elastodynamic problems. Cruise [29] extended the numerical formulation of boundary integral equations to solve problems in 3-D elastostatics. Jawson and Maiti [30], Newton and Tottenham [31] and Forbes and Robinson [32] presented integral equation formulations for elastic plate and shell problems.

Harrington et al. [33] applied the indirect integral equation approach to solve problems in electromagnetics governed by Laplace's equation. Butterfield and Banerjee [34, 35] also applied the indirect integral equation method to the geotechnical problem of pile foundation. During the years 1970–1972, the application of the integral equation method was extended to other areas of engineering science, such as transient heat conduction problems and linear viscoelasticity theory by Rizzo and Shippy [36, 37], fracture mechanics by Cruise and Van Buren [38], plasticity by Swedlow and Cruise [39], water wave scattering problems by Shaw [40] and Lee [41], infinite-domain problems in electromagnetics by Silvester and Hsieh [42] and McDonald and Wexler [43], and orthotopic elasticity problems by Benjumea and Sikarskie [44].

In 1973, Cruise [45] first used the term **BIEM** in the context of 3-D stress analysis with the "direct" method. In the years 1973–1977, both direct and indirect versions of the integral equation method were used to solve problems in elasticity [46–48], torsion [49], fracture [50, 51], plasticity [49, 52, 53], viscous fluid flow [54, 55], ground water flow [56, 57], and thermoelasticity [58]. The first book on the application of BIEM, which was really a collection of articles edited by Cruise and Rizzo [59], was published in 1975.

Banerjee and Butterfield [60] and Brebbia and Dominguez [61] first used the term **BEM** when they recognized the possibility of generalizing discretizations of the boundary problem. Brebbia, together with Dominguez [61–63], first formulated boundary element equations using weighted residual method (WRM) and showed that many numerical method formulations including BEM and FEM can be obtained as special cases of general WRM. This proof provided a connection between the BEM and other numerical techniques like the FEM. The first textbook in boundary integral method was written by Jawson and Symm [64] in 1977. The following year Brebbia [63] published the second textbook on the BEM. Both books covered the application of the BEM to potential theory and theory of elasticity. Zienkiewicz et al. [65, 66] and Atluri and Grannell [67] also showed the connection between the BEM and the FEM using variational principles, and presented techniques for combining the two methods. At about the same time, Brebbia and Butterfield [68] demonstrated the formal equivalence of direct and indirect BEMs.

The research and publication on the BEM increased dramatically in early 1980s and spread into numerous fields of engineering science. In 1980, Brebbia and Walker [69] rewrote the book published two years earlier [63] in an expanded form by adding one chapter on nonlinear and time-dependent problems and another chapter on zoning, approximate boundary elements and combination of the BEM and the FEM. The first comprehensive book on the BEM was published by Banerjee and Butterfield [70] in 1981, followed by Brebbia, Telles and Wrobel [71] in 1984. In the same period, a number of books were published on special topics, e.g., on creep and fracture by Mukherjee [72], on elasticity by Parton and Perlin [73], on solid mechanics by Crouch and Starfield [74], on porous media flow by Liggett and Liu [75], on inelastic problems by Telles [76], on geomechanics by Venturini [77], on complex variable method potential theory by Hromadka II [78], and on potential theory by Ingham and Kelmanson [79].

In addition to numerous research articles published every year in different journals, occasional books are being published which are collections of articles contributed by experts on BEMs in specialized fields [80–88]. Also, regular conferences for the presentation of research papers on BEMs are held every year throughout the world, and conference proceedings are published [89–93]. A journal entitled *Engineering Analysis with Boundary Elements*, fully dedicated to publishing research findings on the BEMs is published regularly under the editorship of Brebbia, Shaw, Tanaka and Aliabadi [94]. A companion communication, *Boundary Elements Communications*, publishes short technical notes on the BEM and lists books and research articles published elsewhere [95]. Two technical societies, ISBE (International Society for Boundary Elements) and IABEM (International Association of Boundary Element Methods), are involved in activities related to boundary element research, education and publication. A few large-scale computer programs, such as, BEASY [96], BEST3D [97], GPBEST [98], SYSNOISE [99], BEMAP [100] and COMET/BEA [101], have been developed and are used by a cross section of engineers.

1.4. History of boundary element eigenformulations

The BEM formulations use the free-space Green's functions as the "test" or "weighting" functions, which are usually transcendental. The implication is that the algebraic eigenvalue formulation in the BEM cannot be posed in a straightforward manner, as the frequency parameters are implicitly embedded in the kernel functions. Consequently, early attempts of BEM eigenvalue formulations were confined to using the frequency sweep method or the determinant search method (DSM) [102–114]. In 1974, Vivoli and Filippi [102] used the DSM to compute acoustic resonant frequencies. The Green's function in this case is complex, and frequency search is conducted on the complex matrix. However, it is possible to employ arbitrary singular solutions with real variables as the fundamental solutions, which would lead to real matrices for determinant search. In 1976, DeMay [103, 104] used this approach to calculate resonant frequencies of Helmholtz equations. The DSM was also used by Hutchinson [105], Hutchinson and Wong [106], Wong and Hutchinson [107], Tai and Shaw [108], Shaw [109], Niwa et al. [110], Hutchinson [111, 112], Adeye et al. [113], and Zhou [114] for Helmholtz equations, plate problems, and membrane vibrations.

In 1980, Bezine [115], in an attempt to set up the algebraic eigenproblem, treated the "inertia" term, containing frequency parameter in it, separately from the remaining term(s) of the governing differential equation for eigenvalue analysis. A simpler fundamental solution, free from the frequency parameter, was used to convert the latter term(s) into a stiffness-type matrix through the boundary discretization. Bezine then divided the domain into internal cells, in addition to the boundary discretization, used shape functions to interpolate the dependent variable in the inertia term, and performed integration of the fundamental solution and the shape function on the domain cells to obtain an additional matrix. After the application of appropriate boundary conditions, the matrices were cast into an algebraic eigenvalue problem. Bezine used this method to solve plate vibration problems. This procedure, based on both boundary and domain discretizations, is designated as the internal cell method (ICM).

In 1982, Nardini and Brebbia [116], like Bezine [115], treated the inertia term separately. However, rather than discretizing the domain, they approximated the dependent variable, contained in the inertia term, by a set of global shape functions and applied the divergence theorem to the term. Thus, the domain integral was converted to the boundary. Thus, Nardini and Brebbia were the researchers who formulated the first boundary-only algebraic eigenvalue problem in the context of BEM. This procedure was first implemented in elastodynamics [116–119] to set up the algebraic eigenproblem. Since, in the technique, the divergence theorem is applied twice, the method was later given the name "Dual Reciprocity Method" (DRM) [120]. Nardini and Brebbia [116] and Partridge and Brebbia [121] suggested a few variations of the global shape functions approximating the dependent variable in the inertial term and the need for adding internal degrees of freedom to improve the accuracy in the computation of the inertial term.

Kanarachos and Provatidis [122] used an indirect formulation to set up the algebraic acoustic eigenvalue problem and showed that the BEM mass matrix must be computed on the basis of a complete functional set, which forces the introduction of source points inside the domain in addition to the boundary collocation points. They also showed that the approximate boundary functions used by Nardini and Brebbia [116] represent only first-order approximations of the "exact functions," designated as the "Poisson-adjusted" functions, presented by them.

Ahmad and Banerjee [123] proposed a slightly different method, which they called Particular Integral Method (PIM), of formulating the generalized eigenvalue problem using the BEM, and applied the method to solve eigenvalue problems in 2-D elasticity. Banerjee et al. [124] applied the PIM to formulate generalized eigenvalue problem in acoustics. In this method, the pressure amplitude is considered to be composed of two components, a complementary function and a particular solution. Wang and Banerjee [125, 126] used PIM to perform axisymmetric as well as non-axisymmetric free-vibration analyses of axisymmetric elastic bodies, and Wilson et al. [127] used it for the free-vibration analysis of 3-D elastic solids. Agnantiaris et al. [128, 129] later applied DRM to analyze free and forced vibration problems of 3-D, non-axisymmetric and axisymmetric 3-D elastic solids. Their study showed that the use of higher order radial basis functions in the evaluation of the inertia term did not noticeably affect the quality of the solution. The DRM was also employed to solve for the free vibration problems of 3-D anisotropic solids [130]. The authors here used a certain number of internal collocation points to accurately compute the mass matrix.

Ali et al. [131] and Rajakumar et al. [132] pointed out that the acoustic eigenvalue problems, especially the most important class with acoustically hard boundaries, can be formulated in terms of fictitious density function, instead of physical variable, thereby avoiding inversion of a large matrix. Ali et al. [131] also brought out the subtle distinction between the free vibration problems in elasticity and acoustic eigenfrequency analyses. They observed that the mode shapes in the former case are conditioned solely by the boundary of the domain, whereas those in the latter case are governed not only by the boundary conditions, but also by the continuity conditions of the eigenfunctions in the domain. As a consequence, an accurate acoustic eigenfrequency analysis of chunky-shaped acoustic cavities may require additional internal collocation points or zoned boundary elements.

Coyette and Fyfe [133] also formulated the acoustic eigenvalue problem in terms of the fictitious function, rather than the pressure amplitude, thereby avoiding a matrix inversion. Bialecki et al. [134] later extended the method to solve transient heat conduction problems with arbitrary sets of boundary conditions. They also pointed out the applicability of the method to differential equations governing diffusion, wave propagation and similar physical phenomena.

In 1992, Raveendra and Banerjee [135] performed acoustic eigenvalue analysis by utilizing complete polynomial-based functions to approximate the pressure in the inertia term. The use of piece-wise polynomials, as opposed to global interpolation functions, to approximate the field pressure amplitude, did not, however, improve the accuracy of eigenfrequencies. Rajakumar and Ali [136] formulated damped acoustic boundary element eigenproblems including sound absorption at the boundary. Note that the eigenformulation in this case led to a quadratic eigenproblem. Rajakumar et al. [137] presented a coupled eigenvalue formulation for fluid–structure systems in which the enclosed fluid was modeled using boundary elements and the structure using finite elements.

Nowak [138] and Nowak and Brebbia [139] proposed the Multiple Reciprocity Method (MRM), in which Gauss's divergence theorem is repeatedly applied to the domain integral term using higher order Green's functions until the domain term becomes negligible. Nowak and Brebbia [140] later applied the method to the Helmholtz equation. Kamiya and Andoh [141] applied the MRM to acoustic eigenvalue problem and solved for resonant frequencies using Newton–Raphson iteration along with LU decomposition. Kamiya and Andoh [142] used a simple matrix augmentation procedure to cast equations into a generalized algebraic eigenproblem. Now the problem could be solved using generalized eigensolvers. In a paper published in 1993, Kamiya et al. [143] provided a good review of the boundary element eigenvalue formulations currently available in the literature with a special emphasis on acoustic eigenanalysis.

Kirkup and Amini [144] proposed the Series Expansion Method (SEM), in which the eigenformulation equation of the DSM was expanded into a series in frequency parameter. Kamiya et al. [143] showed that this series equation (real part) is equivalent to the equation derived using the MRM. A matrix augmentation procedure was then be used to set up the algebraic generalized eigenvalue problem. In 1994, Polyzos et al. [145] showed that the DRM and the PIM are equivalent approaches for treating domain integral terms in the BEM.

Davies and Moslehy [146] used DRM to determine the natural frequencies and mode shapes of thin elastic plates. They inserted additional internal nodes in the domain and employed simpler forms of radial approximating functions in evaluating the inertia term. Davies and Moslehy observed that the accuracy of the eigensolution of the thin plates did not improve appreciably with the use of more complicated forms of the approximating functions. Kamiya et al. [147] employed an h-version of the adaptive mesh refinement technique for the first time in conjunction MRM and Newton iteration to accurately compute acoustic resonant frequencies by BEM.

The boundary element eigenvalue formulations, discussed so far, produce unsymmetric and non-positive definite mass and stiffness matrices. Davì and Milazzo [148] developed a mixed variational principle in which they expressed the functional in terms of independent domain and boundary variables. They employed non-singular static fundamental solutions. DRM-type reciprocity theorem was used to transform the inertia term into boundary-only integrals. Their process resulted into symmetric and positive definite mass and stiffness matrices. Indirect Trefftz method has also been proposed to arrive at symmetric system matrices for the linear algebraic eigenvalue problem [149]. The generalized singular-value decomposition and Tikhonov's regularization methods were employed here in order to overcome the difficulties of spurious eigensolutions and numerical instability associated with indirect Trefftz method.

Niku and Adey [150] observed that the computational costs associated with DRM formulations are relatively high. They considered the diagonalization of the mass and associated matrices in order to reduce the mathematical operation count. They however admitted that it would be necessary to find mathematical justification for such diagonalization.

Chen and Wong [151] combined conventional MRM formulation with the hypersingular equation of DRM to analytically derive eigensolutions for one-dimensional problems. This combined method was later given the name dual MRM and was applied, to determine the natural frequencies and natural modes of an Euler beam [152], a rod [153] as well as square, rectangular and circular and acoustic cavities [154–157]. The single value decomposition method was employed to remove spurious modes.

Ingber et al. [158] found that the direct domain integration technique (ICM), especially with multipole acceleration, can evaluate the inertia term more efficiently than DRM or PIM in terms of CPU cost, memory requirements and accuracy of eigensolution. They remarked that the ICM may be more efficient than DRM/PIM even though the former requires domain discretization. This is because advanced preprocessors have become readily available in recent years.

1.5. Organization of the book

This book is intended to be self-contained. The relevant theories required for a complete understanding of boundary element eigenvalue analysis are provided in the book. The fundamentals of the BEM are presented in Chapters 2 through 4 using the potential problem as an example. Chapter 2 not only presents the essentials of the boundary element formulation, but it also describes a variety of other methods of formulating boundary element equations. Methods that are thought to be in contrast to each other, for example, direct and indirect formulations, weighted residuals, and Green's integral theorem methods, are presented in this chapter.

Isoparametric higher order boundary element formulations in 2-D and 3-D are covered in Chapter 3. The ways of dealing with anisotropic media and axisymmetric bodies are shown in Chapter 4. Although typically boundary element formulations produce full matrices, this chapter shows the so-called zoning technique by which banded system matrices can be produced.

This is followed by the application of the BEM to the time-harmonic analysis in elasticity and acoustics. The relevant theories of elasticity and acoustics are also presented in Chapter 5. Chapter 6 contrasts boundary element formulations with finite element formulations in solving dynamic problems in acoustics and elasticity. The concept of using so-called static fundamental solutions in solving dynamic problems is introduced here, as it is central to formulating algebraic eigenvalue problems using the BEM. The essentials of setting up non-algebraic eigenvalue equations, i.e., characteristic equations, are also presented in Chapter 6.

An algebraic eigenvalue formulation based on combined BEMs and FEMs is presented in Chapter 7, and it is then applied to plate vibration problems. The formulation is designated as the ICM. The ICM is not a boundary-only approach; complete boundary-only algebraic boundary element eigenvalue formulations are presented in Chapters 8 through 10. The principal boundary element algebraic eigenvalue formulations, such as the DRM and the PIM, are developed in Chapter 8. These formulations utilize an associated "static fundamental solution," as opposed to a usual fundamental solution, and employ extra integral transformations in addition to those already required to formulate regular boundary element equations. A few variations of the DRM and the PIM, such as the MRM, polynomial-based PIM, etc., are described in Chapter 9. In Chapter 10 methods are developed that allow us to compute resonant frequencies of fluid enclosed by vibrating or absorbing boundary structure.

The BEM typically produces unsymmetric and full system matrices, which require unsymmetric eigensolvers for their solution. Chapter 11 develops Lanczos-based eigensolvers for unsymmetric system matrices. Both linear and quadratic unsymmetric eigensolvers are presented in Chapter 11. In Chapter 12 we compare boundary element eigenformulations with those in FEM. The shortcomings of the current boundary element eigenvalue formulations are pointed out, along with future research possibilities in this subject. Finally, all the references cited in the book are given.

Chapter 2

Boundary Element Method Fundamentals

2.1. Introduction

For an understanding of the boundary element eigenvalue formulations to be developed in the subsequent chapters, a working knowledge of the fundamentals of the boundary element method (BEM) is essential. This chapter is dedicated to introducing the BEM to the reader. Although our objective in the book is to develop numerical techniques for the computation of resonant frequencies in *acoustics* and *elasticity*, we shall present the basic principles of the BEM using *potential problems* as an illustration. This is because potential problems:

(a) Can be represented by a simple scalar unknown variable;
(b) Are governed by a relatively simple governing equation, e.g., the Laplace's equation; and
(c) Represent a broad class of physical phenomena, e.g., heat conduction, potential flow, seepage, magnetic potential, electrostatics, torsion of shafts, corrosion and many others.

The weighted residual technique is used as the main vehicle to formulate the integral equations, although the classical technique that makes use of the Green's integral transformation identities is also touched upon. Furthermore, the so-called direct boundary element technique is used throughout the book. However, a brief summary of the essentials of the indirect BEM is provided in this chapter. The fundamentals of the application of BEM in the fields of elasticity and acoustics are covered in later chapters.

As mentioned above, potential problems are governed by the Laplace's equation. Consider an arbitrary domain Ω bounded by a surface Γ, as shown in Figure 2.1. We denote a source point and a field point inside the domain Ω by "p" and "q" respectively and the corresponding points on the boundary Γ by "P" and "Q". Let $u(q)$ be the potential function defined in the domain Ω. The boundary value problem can be defined as:

$$\nabla^2 u = 0 \quad \text{in}\,\Omega \tag{2.1}$$

(a) $u = u_{\text{b}}$ on one part of the boundary Γ_u (Dirichlet boundary condition) and
(b) $v\,(=\partial u/\partial n) = v_{\text{b}}$ on the rest of the boundary Γ_v (Neumann boundary condition).

Thus, $\Gamma_u + \Gamma_v = \Gamma$. n is the outward normal to the boundary Γ.

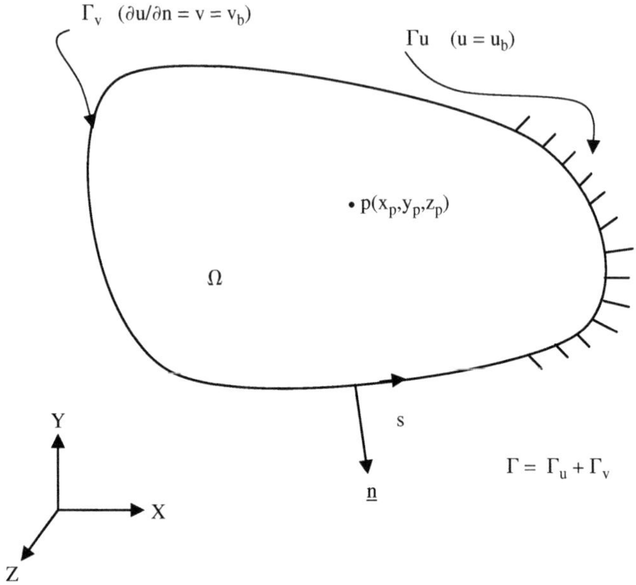

Figure 2.1. Arbitrary domain for potential problem.

The boundary value problem can be discretized in BEM using several different approaches. The main classification would fall into two broad categories: direct method and indirect method.

2.2. Direct method: weighted residuals

In this section, we shall develop the boundary element formulation using the direct method employing weighted residuals technique. The weighted residual method is widely used because of its appeal to a wider audience in computational mechanics. The boundary element formulation can also be developed by another direct method which employs Green's integral identity. We shall present the Green identity-based direct boundary element formulation in Section 2.4.

2.2.1. Weighted residual statements

Let $u^*(p, q)$ be a weighting function. The meaning of u^* will become clear later. The arguments of the functions are omitted in the subsequent developments in order to preserve the simplicity of the presentation. They will be brought back whenever there is a need to distinguish between an internal point and a boundary point or between a field point and a source point.

Employing the weighted residual principle of minimizing the error in solutions of u and v, a weak form of the boundary value problem [eqn. (2.1)] can now be written in the following fashion:

$$\int_{\Omega} (\nabla^2 u)\, u^* \mathrm{d}\Omega = \int_{\Gamma_v} (v - v_\mathrm{b})\, u^* \mathrm{d}\Gamma - \int_{\Gamma_u} (u - u_\mathrm{b})\, v^* \mathrm{d}\Gamma \tag{2.2}$$

where $v^* = \partial u^*/\partial n$. In order to develop the formulation, we will need to integrate the left-hand side of this equation by parts. This will require the use of the Green's identity, which can be written as:

$$\int_\Omega (\nabla^2 u)\, u^* \mathrm{d}\Omega = \int_\Gamma \frac{\partial u}{\partial n} u^* \mathrm{d}\Gamma - \int_\Omega (\underline{\nabla} u \cdot \underline{\nabla} u^*)\mathrm{d}\Omega \tag{2.3}$$

Applying this identity to equation (2.2) and recognizing the fact that $\Gamma_u + \Gamma_v = \Gamma$, one obtains:

$$-\int_\Omega (\underline{\nabla} u \cdot \underline{\nabla} u^*)\mathrm{d}\Omega = -\int_{\Gamma_v} v_b u^* \mathrm{d}\Gamma - \int_{\Gamma_u} v u^* \mathrm{d}\Gamma - \int_{\Gamma_u} u v^* \mathrm{d}\Gamma + \int_{\Gamma_u} u_b v^* \mathrm{d}\Gamma \tag{2.4}$$

Applying the identity one more time to this equation,

$$\int_\Omega (\nabla^2 u^*) u \mathrm{d}\Omega = -\int_{\Gamma_v} v_b u^* \mathrm{d}\Gamma - \int_{\Gamma_u} v u^* \mathrm{d}\Gamma + \int_{\Gamma_v} u v^* \mathrm{d}\Gamma + \int_{\Gamma_u} u_b v^* \mathrm{d}\Gamma \tag{2.5}$$

Note that the finite element formulation of the Laplace's equation stops at equation (2.4). The term $(\underline{\nabla} u \cdot \underline{\nabla} u^*)$ ensures symmetry of the coefficient matrices. On the contrary, in equation (2.5), which is the basic boundary element equation, the Laplacian operator has got completely shifted from the function u to the weighting function u^*. Also, the BEM utilizes a special form of weighting function, called the free-space Green's function. The Green's function is designated as the *fundamental solution* in the boundary element literature. Green's function is the solution to a given differential equation due to a point source placed in a domain of infinite extent. Therefore, for the Laplace's equation at hand the Green's function can be obtained by solving the following equation:

$$\nabla^2 u^*(p,q) + \delta(p,q) = 0 \tag{2.6}$$

$\delta(p,q)$ is the Dirac delta which is infinity at the point p and zero elsewhere and has the property: $\int_\Omega \delta(p,q) = 1$. Also, Dirac delta has a "picking" property such that for any function $f(q)$:

$$\int f(q)\, \delta(p,q) = f(p) \tag{2.7}$$

The fundamental solution for equation (2.6), i.e., the Green's function for the Laplace's equation is given by:

$$u^*(p,q) = \frac{1}{2\pi} \ln \frac{1}{r(p,q)} \qquad \text{(2-D)} \tag{2.8}$$

$$u^*(p,q) = \frac{1}{4\pi} \frac{1}{r(p,q)} \qquad \text{(3-D)} \tag{2.9}$$

$r(p,q)$ is the distance between the source point p and the field point or observation point q. Substituting equation (2.6) into equation (2.5) and utilizing the property of equation (2.7), we arrive at the following boundary integral statement:

$$u(p) + \int_{\Gamma_v} u v^* \mathrm{d}\Gamma + \int_{\Gamma_u} u_b v^* \mathrm{d}\Gamma = \int_{\Gamma_v} v_b u^* \mathrm{d}\Gamma + \int_{\Gamma_u} v u^* \mathrm{d}\Gamma \tag{2.10}$$

This is an integral equation, which is yet another form of the weighted residual statement that we started with. It forms the starting point for the boundary element formulation. It is worth pointing out that equation (2.4), which is the basis for the finite element formulation, consists of integrals over the domain Ω. In contrast, equation (2.10) contains integrals over the boundary Γ and a discrete term at any point p in the domain, heretofore referred to as the source point. Thus, the boundary element formulation requires integration on the boundary alone.

Note that equation (2.10) calculates the value of the function u at any point p within the domain. However, this cannot yet be used to evaluate u at the boundary of the domain because the Green's function that forms part of the integrands is singular on the boundary.

2.2.2. Development of boundary integral equation

In order to develop a numerical technique that leads to the discretization of only the boundary, equation (2.10) of the previous section needs to be evaluated at the boundary Γ. However, it cannot be achieved in its present form because, in that case, the point "p" may coincide with point "q", i.e., the source point may coincide with the field point, thereby yielding $r = 0$. The fundamental solutions given by equations (2.8) and (2.9) are undefined for $r = 0$. The specialization of equation (2.10) to the boundary is, therefore, done through a limiting process.

Consider the portion of the boundary, Γ_v, where Neumann boundary conditions are given and divide Γ_v as $\Gamma_v = \Gamma_{v-\varepsilon} + \Gamma_\varepsilon$ (Fig. 2.2). Γ_ε is a circular arc in 2-D and a spherical surface in 3-D of radius ε centered at P. The first integral term on the left-hand side of equation (2.10) is written as:

$$\int_{\Gamma_v} u \frac{\partial u^*}{\partial n} \mathrm{d}\Gamma = \int_{\Gamma_{v-\varepsilon}} u \frac{\partial u^*}{\partial n} \mathrm{d}\Gamma + \int_{\Gamma_\varepsilon} u \frac{\partial u^*}{\partial n} \mathrm{d}\Gamma \tag{2.11}$$

Consider performing the integration of the last term of this equation. The integration needs to be performed on the boundary Γ_ε. On a circular arc or a spherical surface

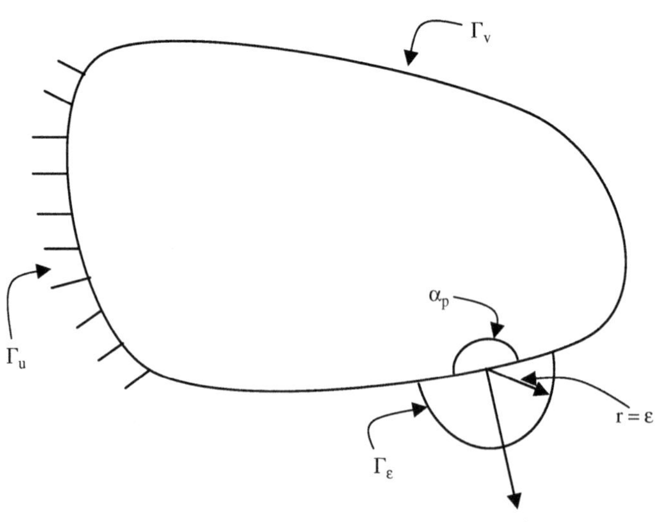

Figure 2.2. The portion of the boundary Γ_v is divided as $\Gamma_v = \Gamma_{v-\varepsilon} + \Gamma_\varepsilon$.

$\partial u^*/\partial n = \partial u^*/\partial r$. Thus,

$$\lim_{\varepsilon\to 0}\left[\int_{\Gamma_\varepsilon} u\frac{\partial u^*}{\partial n}\mathrm{d}\Gamma\right] = \lim_{\varepsilon\to 0}\left[\int_{\Gamma_\varepsilon} u\frac{\partial u^*}{\partial r}\mathrm{d}\Gamma\right] = \lim_{\varepsilon\to 0}\left[-\int_{\Gamma_\varepsilon} u\frac{1}{2\pi\varepsilon}\mathrm{d}\Gamma\right] \quad \text{(2-D)} \tag{2.12a}$$

$$\lim_{\varepsilon\to 0}\left[\int_{\Gamma_\varepsilon} u\frac{\partial u^*}{\partial r}\mathrm{d}\Gamma\right] = \lim_{\varepsilon\to 0}\left[-\int_{\Gamma_\varepsilon} u\frac{1}{4\pi\varepsilon^2}\mathrm{d}\Gamma\right] \quad \text{(3-D)} \tag{2.12b}$$

Let us assume for now that the source point P under consideration is located on a straight (smooth) boundary segment. Remember that the upper case "P" is our notation for a source point on the boundary. The limit in the above integration can be evaluated as follows $[u_i = u(P)]$:

$$\lim_{\varepsilon\to 0}\left[-\int_{\Gamma_\varepsilon} u\frac{1}{2\pi\varepsilon}\mathrm{d}\Gamma\right] = \lim_{\varepsilon\to 0}\left[\int_{\Gamma_\varepsilon} -u_i\frac{\pi\varepsilon}{2\pi\varepsilon}\right] = -\frac{u_i}{2} \quad \text{(2-D)} \tag{2.13a}$$

$$\lim_{\varepsilon\to 0}\left[-\int_{\Gamma_\varepsilon} u\frac{1}{4\pi\varepsilon^2}\mathrm{d}\Gamma\right] = \lim_{\varepsilon\to 0}\left[\int_{\Gamma_\varepsilon} -u_i\frac{2\pi\varepsilon^2}{4\pi\varepsilon^2}\right] = -\frac{u_i}{2} \quad \text{(3-D)} \tag{2.13b}$$

The other integral term in equation (2.10) to be evaluated on the Γ_v boundary can be dealt with in similar fashion:

$$\int_{\Gamma_v} vu^*\mathrm{d}\Gamma = \int_{\Gamma_{v-\varepsilon}} vu^*\mathrm{d}\Gamma + \int_{\Gamma_\varepsilon} vu^*\mathrm{d}\Gamma \tag{2.14}$$

Now, taking the limit on the boundary Γ_ε,

$$\lim_{\varepsilon\to 0}\left[\int_{\Gamma_\varepsilon} v\frac{1}{2\pi}\ln\left(\frac{1}{\varepsilon}\right)\mathrm{d}\Gamma\right] = \lim_{\varepsilon\to 0}\left[v_i\frac{1}{2\pi}\ln\left(\frac{1}{\varepsilon}\right)(\pi\varepsilon)\right] = 0 \quad \text{(2-D)} \tag{2.15a}$$

$$\lim_{\varepsilon\to 0}\left[\int_{\Gamma_\varepsilon} v\frac{1}{4\pi\varepsilon}\mathrm{d}\Gamma\right] = \lim_{\varepsilon\to 0}\left[v_i\frac{1}{4\pi\varepsilon}(2\pi\varepsilon^2)\right] = 0 \quad \text{(3-D)} \tag{2.15b}$$

We obtained the result in the 2-D case (eqn. 2.15a) by applying L'Hospital's rule:

$$\lim_{\varepsilon\to 0}\left[\frac{\ln\varepsilon}{\left(\dfrac{1}{\varepsilon}\right)}\right] = 0 \quad \text{(2-D)} \tag{2.16}$$

Also, as $\varepsilon \to 0$, $\Gamma_{v-\varepsilon} \to \Gamma_v$. Thus, equation (2.10) becomes:

$$\frac{1}{2}u(P) + \int_{\Gamma_v} uv^*\mathrm{d}\Gamma + \int_{\Gamma_u} u_\mathrm{b}v^*\mathrm{d}\Gamma = \int_{\Gamma_v} v_\mathrm{b}\,u^*\mathrm{d}\Gamma + \int_{\Gamma_u} vu^*\mathrm{d}\Gamma \quad \text{(2-D) \& (3-D)} \tag{2.17}$$

Note that the limit of the equation was taken on the Γ_v boundary. The result would be the same if it were performed on the Γ_u boundary. However, in taking an interior domain point to the boundary, we would either arrive at the Γ_u boundary or at the Γ_v boundary and not both. Without making any distinction between the specified and the unknown quantities and making use of the fact that $\Gamma = \Gamma_v + \Gamma_u$, the above equation, valid on boundary points, can be written as:

$$\frac{1}{2}u(P) + \int_{\Gamma} uv^* \mathrm{d}\Gamma = \int_{\Gamma} vu^* \mathrm{d}\Gamma \tag{2.18}$$

In order to evaluate the limit in equation (2.13) it was assumed that the boundary at the point P was smooth which led to the boundary element equation (2.18). In case the boundary at that point is not smooth, this equation is written as:

$$C_P u(P) + \int_{\Gamma} uv^* \mathrm{d}\Gamma = \int_{\Gamma} vu^* \mathrm{d}\Gamma \tag{2.19}$$

where C_P is a coefficient to be evaluated at the boundary point P. In the case of 2-D, it is easy to visualize C_P as the ratio of the external angle and 2π, i.e., $C_P = (2\pi - \alpha_P)/2\pi$, where α_P is the internal angle. In actual discretization, the geometric coefficient C_P is computed through an indirect means without ever requiring to find the angle α_P.

Equation (2.19) is now an entirely boundary-only equation; not only are the integrals performed on the boundary, but all the quantities in the equation are also valid on the boundary. Equation (2.19) is known as the Fredholm integral of the second kind, because the unknown variables are found both inside and outside the integrals. Function value u_P at the source point P is, thus, related to the weighted integrals of the function value u and its derivative $v = \partial u/\partial n$ at the field points around the domain boundary.

Note that the integrals in this equation span the entire boundary of the problem and they are already in place before the boundary has been discretized. Hence, unlike the finite element (FE) method, the boundary integral equation method (BIEM) is known as a global technique and that it produces fully populated matrices. In Section 2.4, we will show the Green's integral theorem approach in deriving the same direct boundary integral equation.

2.2.3. Isoparametric discretization: constant boundary elements

The boundary integral equation (2.18) relates the unknown values of the harmonic function u and its normal derivative v on the boundary Γ. Next step is to break up the boundary curve into small straight segments called "boundary elements" (Fig. 2.3) and assume the unknown values to be constant over each boundary element. Equation (2.18) would then become:

$$\frac{1}{2}u_i + \sum_{j=1}^{N} u_j \int_{\Gamma_j} v_{ij}^* \,\mathrm{d}\Gamma_j = \sum_{j=1}^{N} v_j \int_{\Gamma_j} u_{ij}^* \,\mathrm{d}\Gamma_j \tag{2.20}$$

This equation is written for a source point "i", where "i" varies from 1 to N. The integration is performed on each field element "j" and the results are summed over all the boundary elements (N) in the model including the one that contains the source point "i". Since the unknown quantities u and v are constants over any element "j", they are pulled out of the integration symbol. Note that u and v are discontinuous between any two adjacent elements. The fundamental solution u^* is given by equation (2.8) with $r(p,q) = r_{ij}$. We shall use $r_{ij} = r$ and $\Gamma_j = \Gamma$ for simplicity in the following derivations. The coefficient C_P in equation (2.19) is always equal to ½ for constant elements since

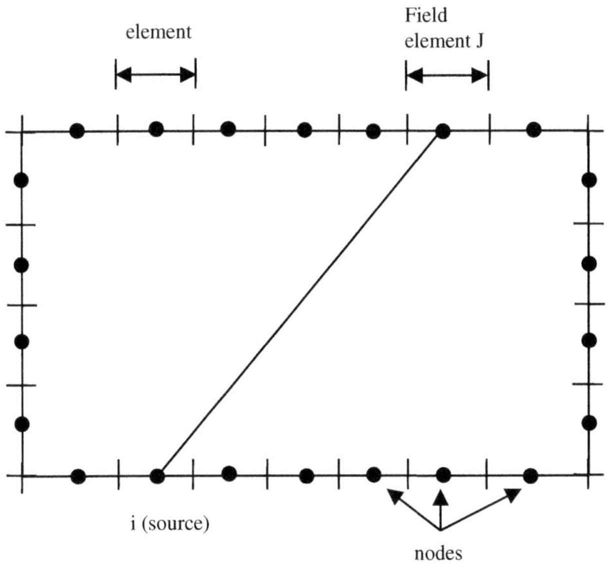

Figure 2.3. Boundary discretization with constant elements.

the angle α_P in Figure 2.2 in this case is 180° or π^c. If we designate the integrated terms by H_{ij} and G_{ij} respectively, equation (2.20) can then be written as:

$$\frac{1}{2}u_i + \sum_{j=1}^{N} u_j \hat{H}_{ij} = \sum_{j=1}^{N} v_j G_{ij} \tag{2.21}$$

The integration of the terms $\hat{H}_{ij}$ is performed as follows:

$$\hat{H}_{ij} = \int_\Gamma \frac{\partial}{\partial n}\left(\frac{1}{2\pi}\ln\frac{1}{r}\right)\mathrm{d}\Gamma \quad \text{(a)}$$

$$= \int_\Gamma \frac{\partial}{\partial r}\left(\frac{1}{2\pi}\ln\frac{1}{r}\right)\frac{\partial r}{\partial n}\mathrm{d}\Gamma \quad \text{(b)}$$

$$= -\int_\Gamma \left(\frac{1}{2\pi r}\right)\frac{\underline{r}\cdot\underline{n}}{|\underline{r}|\,|\underline{n}|}\mathrm{d}\Gamma \quad \text{(c)}$$

$$\hat{H}_{ij} = -\int_\Gamma \frac{\underline{r}\cdot\underline{n}}{2\pi r^2}\mathrm{d}\Gamma, \quad [\because |\underline{n}| = 1] \tag{2.22}$$

The integration on the boundary elements can be divided two categories: one in which the element "j" contains the source point "i" (singular element) and the other in which the element "j" does not contain the source point "i" (non-singular element). The former is so-called because it contains the singular case $r=0$. It is clear from Figure 2.4b that $\underline{r}$ is perpendicular to $\underline{n}$. Hence, $\hat{H}_{ij}=0$ on a singular element ($i=j$). On a non-singular element ($i \neq j$), equation (2.22) can be evaluated using Gaussian quadrature (Fig. 2.4a).

The integration of the terms

$$G_{ij} = \int_\Gamma \frac{1}{2\pi}\ln\frac{1}{r}\mathrm{d}\Gamma \tag{2.23}$$

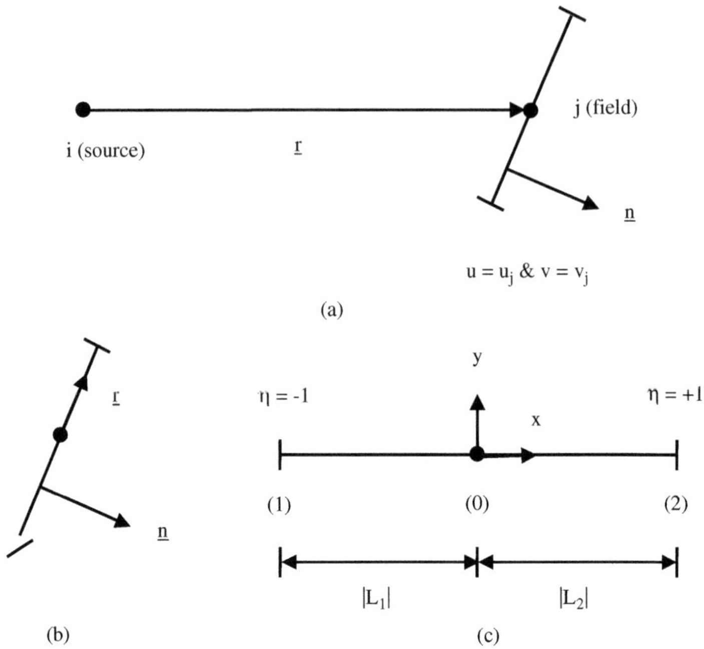

Figure 2.4. Non-singular and singular constant boundary elements. (a) Non-singular boundary element; (b) $\underline{r} \cdot \underline{n} = 0$ on singular boundary element; (c) integration on singular boundary element.

on a non-singular element can be performed using Gaussian quadrature. The integration of this term on a singular element (Fig. 2.4c) can be performed as follows:

$$G_{ij} = \int_\Gamma \frac{1}{2\pi} \ln \frac{1}{r} \mathrm{d}\Gamma = \frac{1}{2\pi} \int_{(1)}^{(2)} \ln \frac{1}{r} \mathrm{d}\Gamma = \frac{1}{\pi} \int_{(0)}^{(2)} \ln \frac{1}{r} \mathrm{d}\Gamma \tag{d}$$

$$= \frac{1}{\pi} |L_1| \left[\ln \frac{1}{|L_1|} + \int_0^1 \ln \frac{1}{\eta} \mathrm{d}\eta \right] \tag{e}$$

$$= \frac{1}{\pi} |L_1| \left[\ln \frac{1}{|L_1|} + 1 \right] \tag{f}$$

If $|L_1| = |L_2| = L/2$, then

$$G_{ij} = \frac{L}{2\pi} \left[\ln\left(\frac{1}{L/2} \right) + 1 \right] \tag{2.24}$$

Equation (2.21) can be finally written in a matrix form:

$$[H]\{u\} = [G]\{v\} \tag{2.25}$$

where:

$$[H] = [\hat{H}] + \pi[I] \tag{2.26}$$

$[H]$ and $[G]$ are $N \times N$ fully populated unsymmetric matrices and $[I]$ is an identity matrix of order N. After applying boundary conditions [eqn. (2.1)], equation (2.25) can be transformed into:

$$[A]\{X\} = \{F\} \tag{2.27}$$

which is a set of N linear equations and can be solved using a linear equation solver. Three types of boundary conditions may arise in practice: (a) pure Dirichlet, (b) pure Neumann and (c) mixed Dirichlet and Neumann. In the first case, the matrix $[G]$ of equation (2.25) will become the final system matrix $[A]$ and the load vector $\{F\}$ will be equal to $[H]\{u\}$. Similarly, for the second case, $[A] = [H]$ and $\{F\} = [G]\{v\}$. For the mixed boundary conditions, the final system matrix and the load vector are formed by transposing all the known boundary conditions on the right-hand side of equation (2.25) through interchange of appropriate columns. The final system matrix $[A]$ once again is a $N \times N$ fully populated and unsymmetric matrix.

With the solution of equation (2.27) the function u and its normal derivative v will be known over the entire boundary Γ. The solution for the function u at any point inside the domain Ω can now be computed using equation (2.10), which, in discretized form, can be written as follows:

$$u_i = \sum_{j=1}^{N} G_{ij} v_j - \sum_{j=1}^{N} \hat{H}_{ij}\, u_j \tag{2.28}$$

where the relation $\Gamma_u + \Gamma_v = \Gamma$ has been used. If desired, the normal derivative v of the potential function can be calculated by differentiating equation (2.10) in the direction of the outward normal n to the boundary Γ and then discretizing it:

$$v_i = \sum_{j=1}^{N} \hat{H}_{ij} v_j - \sum_{j=1}^{N} F_{ij} u_j \tag{2.29}$$

where the integral term F_{ij} is given by:

$$F_{ij} = \int_{\Gamma} \frac{\partial v^*}{\partial n} \mathrm{d}\Gamma \tag{2.30}$$

It was mentioned earlier that Laplace's equation represents a wide variety of problems in engineering science. Two example problems, one in thermal heat conduction and the other in potential fluid flow, are presented below in order to illustrate the use of boundary elements in solving problems governed by Laplace's equation.

2.3. Examples

The following two examples are presented to illustrate the use of the constant BEM developed in the previous section. It may be noted here that unlike in FEM, BEM routinely allows the use of constant shape function to approximate the field variable over the element segment. In BEM formulations, both the field variable and its normal gradient appear as unknown degrees of freedom to be solved. Mathematically, the normal gradient requires a shape function, which is one polynomial order lower than the field variable itself. However, in actual applications, both the field variable and its normal gradient are discretized using equal order shape functions.

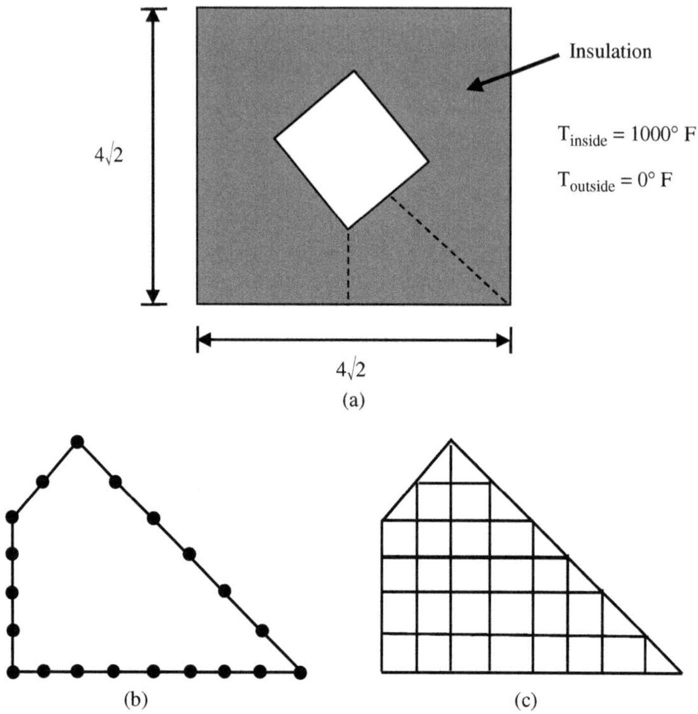

Figure 2.5. (a) An insulated heating duct; the area enclosed in dashed lines is modeled. (b) BE mesh for the duct. (c) FE mesh for the duct.

Table 2.1. Temperature solution at internal points for heating duct.

		Temperature		
x-coordinate	*y*-coordinate	Constant BE	Linear BE	FE
0.3540	0.3540	171.4173	178.6738	176.97
1.0620	1.0620	414.1725	420.8616	417.59
1.0620	0.7080	272.0703	277.7746	276.39
0.7080	1.4160	671.6613	681.3048	675.42
2.1240	0.3540	55.6652	56.5016	56.568

Example 2.1: Heat conduction

Figure 2.5a shows a 2 feet × 2 feet metal heating duct surrounded by insulating materials. The problem is to compute temperature distribution in the insulation material when the duct temperature is maintained at 1000°F and the outside temperature is taken as 0°F. Only one-eighth of the domain is modeled. The boundary element mesh is shown in Figure 2.5b. The problem is solved using constant and linear elements. The formulation for linear element will be presented in the next chapter. Gipson [59] compares his boundary element (BE) results for this problem against a finite difference solution. Here we have performed a finite element analysis of the heating duct maintaining the same level of discretization on the boundary [160]. The FE mesh is shown in Figure 2.5c. All the results are presented in Table 2.1.

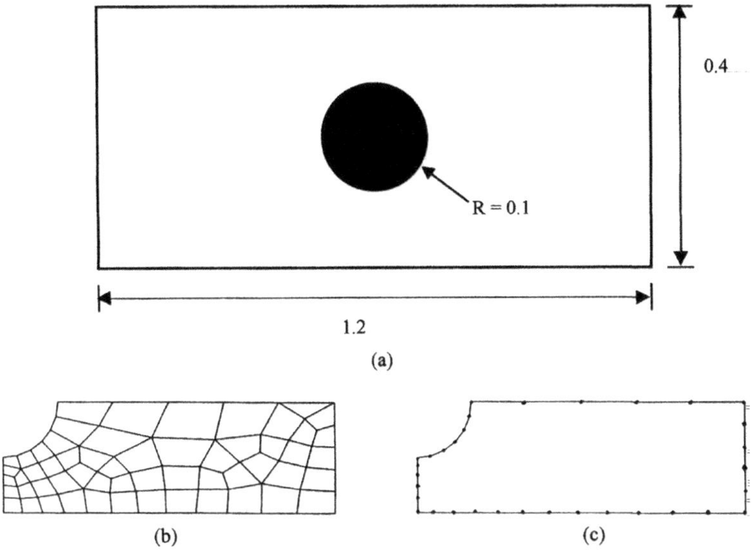

Figure 2.6. (a) Flow around a cylinder between two parallel plates. (b) FE mesh for the flow problem domain. (c) BE mesh for the flow problem domain.

Example 2.2: Potential fluid flow

The BE program written for solving Laplace's equation can be used to solve fluid flow problems by interpreting potential u as streamline function and the potential gradient $v = \partial u/\partial n$ as the velocity along the boundary of the problem domain. For example, consider the problem of fluid flow around a cylinder between two parallel plates, as shown in Figure 2.6a. One quarter of the domain needs to be modeled. The boundary element mesh with linear elements is shown in Figure 2.6c. The potential u, i.e., streamline is taken as zero at the bottom plate and the cylinder surface. The potential gradient $v = \partial u/\partial n$, i.e., the velocity along the vertical boundaries at $x = 0$ and $x = 0.6$ is also zero. The fluid velocity, normal to the boundary at $x = 0$, is assumed to be unity which will result into $u = 2$ at the top plate. The problem is solved by constant and linear boundary elements as well as by the finite elements. The solutions for the streamline functions at the interior points are shown in Table 2.2 for constant as well as linear elements. The mesh used in the finite element analysis is shown in Figure 2.6b. The results from the finite element analysis are also shown in Table 2.2. The BE and FE solutions appear to be in close agreement.

2.4. Direct method: Green's integral theorem

The use of weighted residual technique in formulating boundary integral equations is a relatively recent development [61–63]. Classical approaches utilized Green's integral identities in order to derive boundary integral equations. Let us consider two functions ϕ and ψ defined in the domain Ω of Figure 2.7. Suppose that these functions and their first partial derivatives are continuous in the domain. Green's second integral identity involving these functions and their derivatives can be written as:

$$\int_{\Omega} (\phi \nabla^2 \psi - \psi \nabla^2 \phi)\, \mathrm{d}\Omega = \int_{\Gamma} \left(\phi \frac{\partial \psi}{\partial n} - \psi \frac{\partial \phi}{\partial n} \right) \mathrm{d}\Gamma \tag{2.31}$$

Table 2.2. Streamline solution at interior points for fluid flow.

FE Nodes	x-coord.	y-coord.	Constant BE	Linear BE	Finite element
34	2.16E−01	6.99E−02	6.95E−01	6.95E−01	0.69672
35	1.08E−01	7.07E−02	7.06E−01	7.06E−01	0.70639
36	1.02E−01	1.51E−01	1.51E+00	1.51E+00	1.5067
37	3.24E−01	7.22E−02	6.99E−01	6.99E−01	0.70334
38	1.59E−01	1.40E−01	1.39E+00	1.39E+00	1.3939
39	3.85E−01	9.74E−02	9.10E−01	9.09E−01	0.90688
40	3.66E−01	1.48E−01	1.45E+00	1.45E+00	1.447
41	3.07E−01	1.39E−01	1.38E+00	1.38E+00	1.3755
42	2.33E−01	1.37E−01	1.37E+00	1.37E+00	1.366
43	4.38E−01	1.15E−01	1.01E+00	1.01E+00	1.0118
44	5.03E−01	1.69E−01	1.55E+00	1.55E+00	1.5496
45	4.62E−01	1.63E−01	1.53E+00	1.53E+00	1.5328
46	4.17E−01	1.56E−01	1.50E+00	1.50E+00	1.5013
47	5.25E−02	1.60E−01	1.60E+00	1.60E+00	1.5971
48	6.27E−02	1.06E−01	1.06E+00	1.06E+00	1.0588
49	5.35E−01	1.75E−01	1.59E+00	1.60E+00	1.5881
50	5.43E−01	1.55E−01	1.27E+00	1.28E+00	1.2569
51	5.37E−01	1.08E−01	5.08E−01	5.12E−01	0.49792
52	5.04E−01	9.65E−02	5.52E−01	5.52E−01	0.54674
53	4.82E−01	1.28E−01	1.05E+00	1.05E+00	1.0445
54	4.16E−01	5.49E−02	4.64E−01	4.64E−01	0.47171
55	5.69E−01	1.53E−01	1.17E+00	1.19E+00	1.165
56	5.66E−01	1.33E−01	8.19E−01	8.46E−01	0.8088
57	5.69E−01	1.13E−01	4.05E−01	4.36E−01	0.38894
58	5.67E−01	1.75E−01	1.57E+00	1.58E+00	1.5606
59	3.85E−02	1.28E−01	1.28E+00	1.28E+00	1.2817
60	4.68E−01	7.91E−02	5.48E−01	5.49E−01	0.54814
61	5.26E−01	1.39E−01	1.05E+00	1.05E+00	1.043

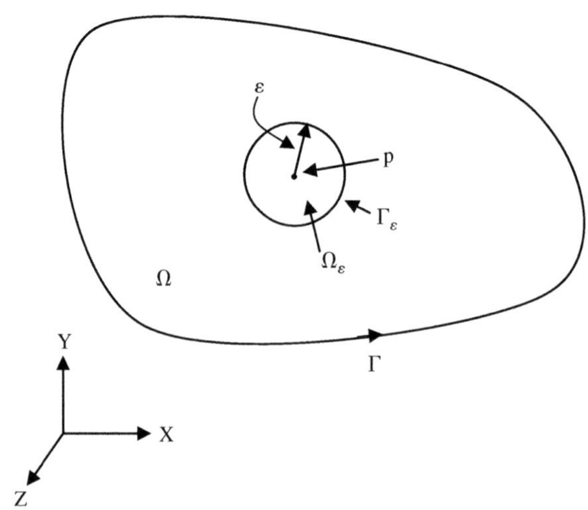

Figure 2.7. Integration over a small circular boundary Γ_ε.

We can identify the functions ϕ and ψ as u and u^* of Section 2.2. Function u is the solution that we seek for the governing Laplace's equation. Therefore, u satisfies the Laplace's equation in the entire domain Ω [eqn. (2.1)]. Since u^* is the fundamental solution to a point source at P, it also satisfies the Laplace's equation for the entire domain Ω, except at the point p [eqn. (2.6)]. Let us isolate this point by putting an arbitrarily small circular region (or a spherical region in 3-D) Ω_ε around the point bounded by $\Gamma_\in$ (Fig. 2.7). Green's second integral identity can be applied to the region Ω–Ω_ε:

$$\int_{\Omega-\Omega_\varepsilon} \left(u\nabla^2 u^* - u^*\nabla^2 u\right) \mathrm{d}\Omega = \int_{\Gamma+\Gamma_\varepsilon} \left(u\frac{\partial u^*}{\partial n} - u^*\frac{\partial u}{\partial n}\right) \mathrm{d}\Gamma \tag{2.32}$$

Note that both boundaries of the region Ω–Ω_ε, viz., Γ and Γ_ε, are included in writing the integral identity. The left-hand side of this equation is identically zero. Let us evaluate the first integral on the right-hand side on the boundary Γ_ε:

$$\operatorname*{Lim}_{\varepsilon\to 0}\left[\int_{\Gamma_\varepsilon} u\frac{\partial u^*}{\partial n}\mathrm{d}\Gamma\right] = \operatorname*{Lim}_{\varepsilon\to 0} -\left[\int_{\Gamma_\varepsilon} u\frac{\partial u^*}{\partial r}\mathrm{d}\Gamma\right] = u_p\left(\frac{2\pi\varepsilon}{2\pi\varepsilon}\right) = u_p \quad \text{(2-D)} \tag{2.33a}$$

$$\operatorname*{Lim}_{\varepsilon\to 0}\left[\int_{\Gamma_\varepsilon} u\frac{\partial u^*}{\partial n}\mathrm{d}\Gamma\right] = \operatorname*{Lim}_{\varepsilon\to 0} -\left[\int_{\Gamma_\varepsilon} u\frac{\partial u^*}{\partial r}\mathrm{d}\Gamma\right] = u_p\left(\frac{4\pi\varepsilon^2}{4\pi\varepsilon^2}\right) = u_p \quad \text{(3-D)} \tag{2.33b}$$

The second integral on the boundary Γ_ε would vanish in the limit as $\varepsilon \to 0$. Thus, equation (2.32) becomes:

$$u_p = \int_\Gamma \left(u^*\frac{\partial u}{\partial n} - u\frac{\partial u^*}{\partial n}\right) \mathrm{d}\Gamma \tag{2.34}$$

This equation states that a harmonic function at a point $p\,(u_p)$ in the domain Ω can be expressed as the sum of a single-layer potential (integral term with the fundamental solution, u^*, in it) with density $\partial u/\partial n$ and a double-layer potential (integral term with the normal derivative of the fundamental solution, $\partial u^*/\partial n$, in it) with density $-u$. We note here that the single-layer potential is continuous, but the double-layer potential experiences a jump as the point p passes through the boundary of the domain. It can be seen that equation (2.34) is essentially identical to equation (2.10) derived using the weighted residual method. The limits for specializing the interior point p to the boundary, leading to equation (2.19), can be taken in the same way as in Section 2.2. We will use weighted residual technique in the rest of the book for deriving boundary element equations.

The materials presented up to this point in this chapter would be adequate for a general understanding of the fundamentals of the boundary element formulation. Very often the approach outlined thus far would be found in boundary element literature that describes the basic methodology. The next section on indirect method, therefore, is presented briefly for the sake of completeness.

2.5. Indirect method

In the direct method, the physical quantities themselves are used as the unknown variables to be solved by numerical means. For example, the harmonic function u and its normal derivative v, defined in Sections 2.2 and 2.4, are solved as unknowns in the direct method. Depending on the physical problem solved, these harmonic functions may represent temperature or velocity of flow or electrical volt. In the so-called semi-direct method, which also uses the direct formulation as derived in Sections 2.2

and 2.4, the unknown function may be taken as the stress function or stream function or magnetic potential function. The physical quantities of the problem at hand can be computed by differentiation of these functions after the unknowns have been solved for using BEM. In the indirect or source method, however, "source" densities are used as the unknowns of the problem. These source densities may or may not have any direct physical meaning for the problem to be solved. The physical quantities are computed using integral expressions in terms of the source densities after the source densities have been solved for.

We have seen from equation (2.34) that a harmonic function at any point in the domain can be expressed as the sum of a single-layer potential with an unknown density and a double-layer potential with another unknown density. Suppose the entire boundary of the problem is of the Dirichlet type, i.e., Γ_u. In that case the unknown harmonic function $u(p)$ may be expressed only by a double-layer potential of unknown density $\sigma(Q)$:

$$u(p) = \int_{\Gamma_u=\Gamma} \sigma(Q) \frac{\partial u^*(p,Q)}{\partial n(Q)} \mathrm{d}\Gamma(Q) \tag{2.35}$$

Since we already know that the double-layer potential experiences a jump as the domain point p approaches the boundary point P, we obtain from equation (2.35) (along the same line of derivation used for eqn. 2.18):

$$u_b(P) = -\frac{1}{2}\sigma(P) + \int_{\Gamma} \sigma(Q) \frac{\partial u^*(P,Q)}{\partial n(Q)} \mathrm{d}\Gamma(Q) \tag{2.36}$$

In equation (2.36) the boundary at the source point P has been assumed to be smooth. This is a Fredholm integral equation of the second kind. Alternatively, Dirichlet boundary value problem can be solved by expressing the unknown harmonic function $u(p)$ only as a single-layer potential with unknown density $\sigma(Q)$:

$$u(p) = \int_{\Gamma_u=\Gamma} \sigma(Q) u^*(p,Q) \mathrm{d}\Gamma(Q) \tag{2.37}$$

As p approaches P, unlike the double-layer potential, the single-layer potential does not experience a jump. Thus, equation (2.37) becomes:

$$u_b(P) = \int_{\Gamma} \sigma(Q) u^*(P,Q) \mathrm{d}\Gamma(Q) \tag{2.38}$$

This is a Fredholm integral equation of the first kind. Equations of this type are more difficult to solve, compared to Fredholm equation of the second kind, because of possible ill-conditioning of the matrices and non-uniqueness of the solution resulting from the discretization of the problem [16].

For the Neumann problem, i.e., if the entire boundary is of the type Γ_v, we can assume that the unknown harmonic function $u(p)$ may be expressed solely as a single-layer potential with unknown density $\sigma(Q)$:

$$u(p) = \int_{\Gamma_v=\Gamma} \sigma(Q) u^*(p,Q) \mathrm{d}\Gamma(Q) \tag{2.39}$$

Taking directional derivative of the function $u(p)$ at a point p in the normal direction $\boldsymbol{n}$ to the boundary we get:

$$\frac{\partial u(p)}{\partial n} = \int_{\Gamma} \sigma(Q) \frac{\partial u^*(p,Q)}{\partial n(p)} \mathrm{d}\Gamma(Q) \tag{2.40}$$

If we take the limit as the internal point p approaches the boundary point P, we obtain the following integral equation:

$$v_b(P) = -\frac{1}{2}\sigma(P) + \int_{\Gamma} \sigma(Q) \frac{\partial u^*(P,Q)}{\partial n(Q)} \mathrm{d}\Gamma(Q) \tag{2.41}$$

This is a Fredholm integral equation of the second kind. Once again, the boundary at the point P is assumed to be smooth. Equation (2.41) will have a solution if the following relation, known as the Gauss condition, is satisfied [60]:

$$\int_{\Gamma} v_b(Q) \mathrm{d}\Gamma(Q) = 0 \tag{2.42}$$

The solution to equation (2.41) is unique only up to an arbitrary additive constant. A unique solution to equation (2.41) can, however, be obtained by imposing some normalization procedure [62].

For the boundary value problem having mixed Dirichlet and Neumann boundary conditions, one can proceed as in the pure Neumann problem and express the function $u(p)$ as a single-layer potential:

$$u(p) = \int_{\Gamma} \sigma(Q) u^*(p,Q) \mathrm{d}\Gamma(Q) + \mathrm{C} \tag{2.43}$$

The constant is added to ensure uniqueness of the solution. Once again, taking a directional derivative, one obtains:

$$v(p) = \int_{\Gamma} \sigma(Q) \frac{\partial u^*(p,Q)}{\partial n(p)} \mathrm{d}\Gamma(Q) \tag{2.44}$$

As the domain point p approaches the boundary point P, equations (2.43) and (2.44) take the form:

$$u_b(P) = \int_{\Gamma_u} \sigma(Q) u^*(P,Q) \mathrm{d}\Gamma(Q) + \mathrm{C} \tag{2.45}$$

$$v_b(P) = -\frac{1}{2}\sigma(P) + \int_{\Gamma_v} \sigma(Q) \frac{\partial u^*(P,Q)}{\partial n(P)} \mathrm{d}\Gamma(Q) \tag{2.46}$$

This equation pair is solved simultaneously for the unknown source density σ on the boundary. The unknown function, u, on the boundary Γ_v and the unknown normal derivative, v, of the function on the boundary Γ_u can then be computed using the following equations:

$$u(P) = \int_{\Gamma_v} \sigma(Q) u^*(P,Q) \mathrm{d}\Gamma(Q) + \mathrm{C} \tag{2.47}$$

$$v(P) = -\frac{1}{2}\sigma(P) + \int_{\Gamma_u} \sigma(Q) \frac{\partial u^*(P,Q)}{\partial n(P)} \mathrm{d}\Gamma(Q) \tag{2.48}$$

2.6. Body forces

The problems that can be solved using boundary element formulation presented in all of the previous sections are driven solely by the boundary conditions. In many practical applications, the domain may contain discrete or distributed sources or body forces, such as, heat generation for heat conduction problem or electrical charges for electrostatic problem. This type of problem is governed by the Poisson's equation:

$$\nabla^2 u = b \qquad \text{in}\ \Omega \tag{2.49}$$

where "b" is the source term. Depending on the type of the source, a number of strategies may be employed to include the effects of the domain term "b".

First, we can transform the boundary-value problem for the Poisson's equation into one for the Laplace's equation by subtracting a particular solution that is independent of the boundary conditions. Suppose we are required to solve the problem: $\nabla^2 u = 4$ with $u = 0$ on Γ. The solution u can be written as a sum of a particular solution u_p and a complementary function u_c: $u = u_\mathrm{p} + u_\mathrm{c}$. $u_\mathrm{p} = x^2 + y^2$ is a particular solution to the problem in 2-D. Thus, the boundary-value problem can be posed in the following fashion with u_c as the unknown function: $\nabla^2 u_\mathrm{c} = 0$ with $u_\mathrm{c} = -u_\mathrm{p}$ on Γ.

Second, for many practical cases it will be difficult to find a particular solution. For example, the values of "b" may be provided in a tabular form applied at a series of points in Ω. In these situations, the boundary element formulation given by equation (2.19) can be extended to include the domain term "b" in the following manner:

$$C_P u + \int_\Gamma u v^* \mathrm{d}\Gamma + \int_\Omega b u^* \mathrm{d}\Omega = \int_\Gamma v u^* \mathrm{d}\Gamma \tag{2.50}$$

The domain term can be computed by dividing the domain Ω into a number of internal cells (Fig. 2.8) and performing numerical integration over these cells. A typical term corresponding to a source point "i" on the boundary Γ can be written as:

$$B_i = \int_\Omega b u^* \mathrm{d}\Omega = \sum_{l=1}^{N_\mathrm{c}} \left[\sum_{k=1}^{N_i} \left\{ \left(b u^*\right)_k w_k \right\} A_l \right] \tag{2.51}$$

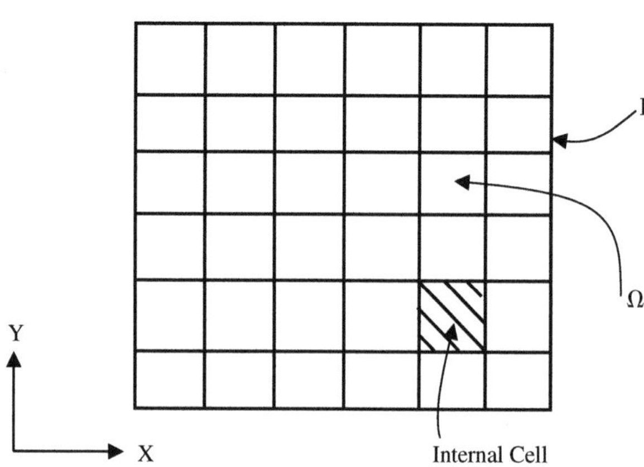

Figure 2.8. Internal cells for domain term integration.

where N_c is the total number of internal cells, N_i is the number of integration points in each cell, w_k are the integration weights and A_l is the area of the "l"th cell. The entire system of equations for the N boundary elements corresponding to equation (2.25) can now be written as:

$$\{B\} + [H]\{u\} = [G]\{v\} \tag{2.52}$$

The discretization of the domain Ω into a number of cells for evaluating the domain term does not introduce any extra unknown into the system of equations. Thus, applying boundary conditions the above set of equations can be rearranged into a system similar to equation (2.27). After this set of equations has been solved, the values of u and v will be known over the entire boundary Γ. The value of the function u at any interior point "i", similar to equation (2.28), can then be computed using the following equation:

$$u_i = \sum_{j=1}^{N} G_{ij} v_j - \sum_{j=1}^{N} \hat{H}_{ij} u_j - B_i \tag{2.53}$$

Third, if the body term "b" is a harmonic function, i.e., if $\nabla^2 b = 0$, then the domain integral term of equation (2.50) can be transformed into equivalent boundary integral terms. To achieve this, we assume that a function U^* exists such that $\nabla^2 U^* = u^*$ and then apply Green's second identity in the following fashion:

$$\int_\Omega \left(b\nabla^2 U^* - U^* \nabla^2 b\right) \mathrm{d}\Omega = \int_\Gamma \left(b\frac{\partial U^*}{\partial n} - U^* \frac{\partial b}{\partial n}\right) \mathrm{d}\Gamma \tag{2.54}$$

Since we assumed that $\nabla^2 b = 0$ and $\nabla^2 U^* = u^*$, we obtain:

$$\int_\Omega b u^* \mathrm{d}\Omega = \int_\Gamma \left(b\frac{\partial U^*}{\partial n} - U^* \frac{\partial b}{\partial n}\right) \mathrm{d}\Gamma \tag{2.55}$$

One form of function U^* that satisfies the relation $\nabla^2 U^* = u^*$ is given by:

$$U^* = \frac{r^2}{8\pi}\left[\ln\left(\frac{1}{r}\right) + 1\right] \tag{2.56}$$

If we substitute the expression (2.55) into equation (2.50), we see that all the terms are now applied on the boundary only.

Fourth, if there are a number of concentrated sources "Q" at discrete interior points of the domain Ω in addition to the distributed sources "b", equation (2.50) can be modified to include the effects of the sources:

$$C_P u + \int_\Gamma u v^* \mathrm{d}\Gamma + \int_\Omega b u^* \mathrm{d}\Omega + \sum \left(Q_k u_k^*\right) = \int_\Gamma v u^* \mathrm{d}\Gamma \tag{2.57}$$

Example 2.3: Twist of a prismatic shaft

The governing differential equation for the twist of a homogeneous prismatic shaft in terms of the stress function u is given as:

$$\nabla^2 u = -2G\theta \tag{2.58}$$

where G is the shear modulus and θ is rate of twist. The shear stresses can be derived from the stress function. The stress function is constant (for convenience it is assumed

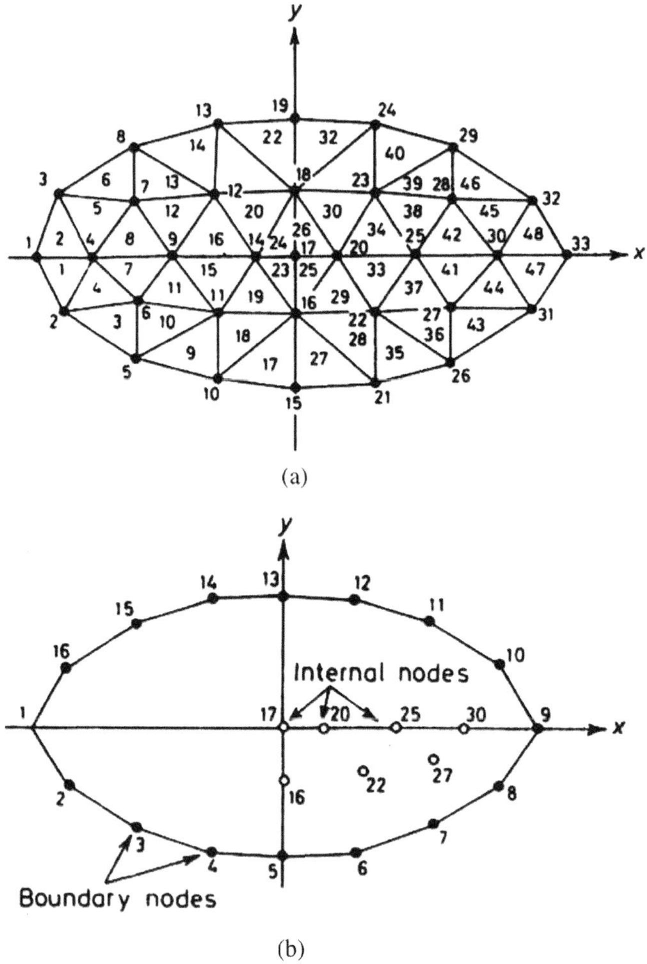

Figure 2.9. (a) FE mesh of the elliptical section of the shaft. (b) BE mesh of the elliptical section.

Table 2.3. Solution of stress function for twist of a shaft.

x-coord	y-coord	Exact solution	Constant BE	Linear BE	Finite element
0.0	0.0	0.8	0.798	0.803	0.793
0.35	0.0	0.7755	0.773	0.778	0.782
1.0	0.0	0.6	0.599	0.602	0.591
1.5	0.0	0.35	0.358	0.351	0.347
0.0	−0.44	0.64512	0.644	0.648	0.681
0.6	−0.44	0.57312	0.571	0.576	0.569
1.2	−0.44	0.35712	0.355	0.359	0.336

to be zero) on the boundary. Let $G = \theta = 1$. Also, let us consider a prismatic shaft of elliptic cross section:

$$\frac{x^2}{a} + \frac{y^2}{b} = 1 \tag{2.59}$$

Let us take $a = 2$ and $b = 1$ for this example [69]. The problem is solved by the boundary element and finite element methods (FEMs). For the finite element analysis, 33 nodes and 48 linear triangular elements are used (Fig. 2.9a) and 16 constant/linear elements are used for the boundary element analysis (Fig. 2.9b). For the BE analysis, the finite elements of Figure 2.9a are used as the internal cells to integrate the domain term. The results of the analyses for a few internal points are presented in Table 2.3. The exact solutions at these internal points are also shown in the table.

Chapter 3

Isoparametric Boundary Elements

3.1. Introduction

The fundamentals of the boundary element method (BEM) were presented in Chapter 2. The reader may notice that if the solution variables are represented as constants over the boundary segments, the developments of Chapter 2 are sufficient for an understanding and implementation of the method. However, for better representation of the geometry as well as better accuracy of the solution variables, often, higher polynomial order representation of the solution variables as well as the geometry is needed. As in the finite element methods (FEM), this leads to isoparametric formulation of the boundary element equations. This chapter will present the basic approach of transforming the global coordinates to normalized local systems and the use of higher order polynomial shape functions in the process of boundary element discretization.

3.2. Two-dimensional linear boundary elements

As in the case of constant boundary element, which was developed in the previous chapter, the boundary curve Γ is once again divided into "N" small straight boundary element segments (Fig. 3.1). Unlike in the constant elements, the unknown function u and its normal derivative v are assumed to vary linearly over each element segment. The nodes are located at the ends of the element segment. It means that adjacent elements share nodes, ensuring inter-element continuity of the function u and its normal derivative v. The angle α_P of Figure 2.2 can no longer be assured to equal to 180°. For the discretized boundary, the equation (2.19) can be written for the i-th source point in the following fashion:

$$C_i u_i + \sum_{j=1}^{N} \int_{\Gamma_j} u v^* \mathrm{d}\Gamma = \sum_{j=1}^{N} \int_{\Gamma_j} v u^* \mathrm{d}\Gamma \tag{3.1}$$

We will shortly show a simple technique to compute the coefficient C_i. The unknown function u and its normal derivative v can no longer be pulled out of the integral sign because they vary linearly over the element segment. A linear boundary element is shown in Figure 3.2 along with the functional variation of u and v. These functions

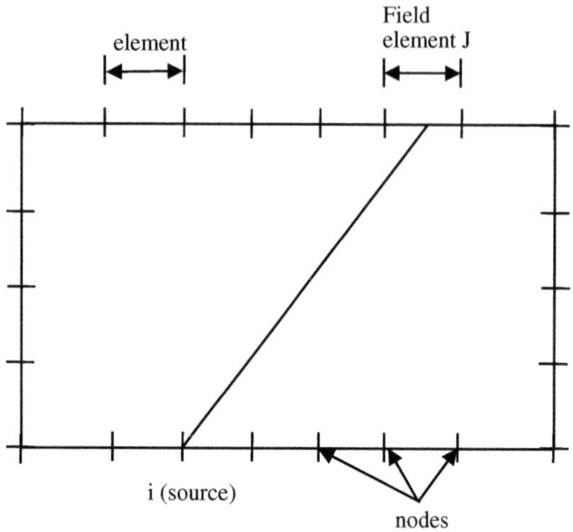

Figure 3.1. Boundary discretization with linear elements.

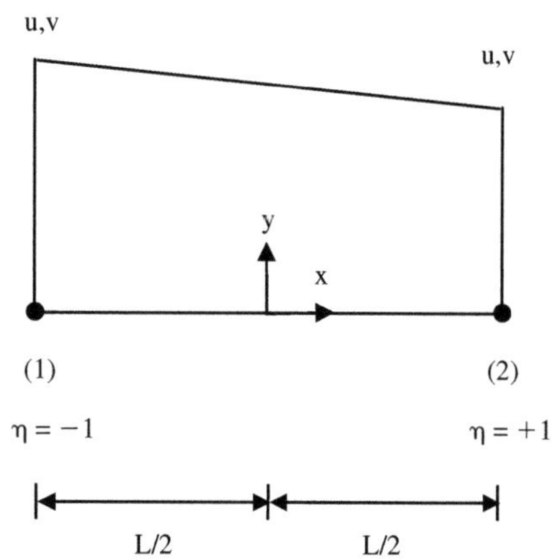

Figure 3.2. A linear boundary element.

may be represented over each element by linear shape functions and nodal values:

$$\left.\begin{aligned} u &= \phi_i u_i \\ v &= \phi_i v_i \end{aligned}\right\} \qquad (i = 1, 2) \tag{3.2}$$

where the shape functions are given by:

$$\left.\begin{aligned} \phi_1(\eta) &= \tfrac{1-\eta}{2} \\ \phi_2(\eta) &= \tfrac{1+\eta}{2} \end{aligned}\right\} \tag{3.3}$$

in which repeated indices imply a summation. The first integral term of equation (3.1) can then be written as:

$$\int_{\Gamma_j} uv^* \, \mathrm{d}\Gamma = \int_{\Gamma_j} [\phi_1 \quad \phi_2] v^* \mathrm{d}\Gamma \begin{Bmatrix} u_1 \\ u_2 \end{Bmatrix} = \begin{bmatrix} \hat{h}_{i1} & \hat{h}_{i2} \end{bmatrix} \begin{Bmatrix} u_1 \\ u_2 \end{Bmatrix} \tag{3.4}$$

with:

$$\left.\begin{aligned}\hat{h}_{i1} &= \int_{\Gamma_j} \phi_1 v^* \, \mathrm{d}\Gamma \\ \hat{h}_{i2} &= \int_{\Gamma_j} \phi_2 v^* \, \mathrm{d}\Gamma\end{aligned}\right\} \tag{3.5}$$

Similarly, the second term of equation (3.1) can be written as:

$$\int_{\Gamma_j} vu^* \, \mathrm{d}\Gamma = \int_{\Gamma_j} [\phi_1 \quad \phi_2] u^* \mathrm{d}\Gamma \begin{Bmatrix} v_1 \\ v_2 \end{Bmatrix} = \begin{bmatrix} g_{i1} & g_{i2} \end{bmatrix} \begin{Bmatrix} v_1 \\ v_2 \end{Bmatrix} \tag{3.6}$$

with:

$$\left.\begin{aligned}g_{i1} &= \int_{\Gamma_j} \phi_1 u^* \, \mathrm{d}\Gamma \\ g_{i2} &= \int_{\Gamma_j} \phi_2 u^* \, \mathrm{d}\Gamma\end{aligned}\right\} \tag{3.7}$$

The assembled equation for the source point "i" can thus be easily written as:

$$C_i u_i + \begin{bmatrix} \hat{H}_{i1} & \hat{H}_{i2} & \cdots & \hat{H}_{iN} \end{bmatrix} \begin{Bmatrix} u_1 \\ u_2 \\ \vdots \\ u_N \end{Bmatrix} = \begin{bmatrix} G_{i1} & G_{i2} & \cdots & G_{iN} \end{bmatrix} \begin{Bmatrix} v_1 \\ v_2 \\ \vdots \\ v_N \end{Bmatrix} \tag{3.8}$$

where, $\hat{H}_{i1} = (\hat{h}_{i1}$ from element "j") + ($\hat{h}_{i2}$ from element "$j-1$"). Similarly, $G_{i1} = (g_{i1}$ from element "j") + (g_{i2} from element "$j-1$"). The equation (3.8) can be written in a concise form as follows:

$$C_i u_i + \sum_{j=1}^{N} \hat{H}_{ij} u_j = \sum_{j=1}^{N} G_{ij} v_j \tag{3.9}$$

In matrix form this will lead to equation (2.25). However, equation (2.26) will now take the form:

$$[H] = [\hat{H}] + [C] \tag{3.10}$$

where $[C]$ is a diagonal matrix whose coefficients are evaluated using a procedure shown below. As in the constant element, we can use Gaussian quadrature for the non-singular elements. Any source point "i" is connected to two adjacent elements "$i-1$" and "i", as shown in Figure 3.3. All four $\hat{h}_{ij}$ terms, two for the $(i-1)$th element and another two for the (i)th element, will equal to zero, as $\boldsymbol{r} \cdot \boldsymbol{n} = 0$. The corresponding g_{ij} terms can be evaluated analytically. The results are recorded below:

$$\left.\begin{aligned}g_{i,i} &= \tfrac{L_i}{2}\left[\tfrac{3}{2} - \ln L_i\right] \\ g_{i,i+1} &= \tfrac{L_i}{2}\left[\tfrac{1}{2} - \ln L_i\right]\end{aligned}\right\} \quad \text{for } (i)\text{th element} \tag{3.11}$$

$$\left.\begin{aligned}g_{i,i} &= \tfrac{L_{i-1}}{2}\left[\tfrac{3}{2} - \ln L_{i-1}\right] \\ g_{i,i-1} &= \tfrac{L_{i-1}}{2}\left[\tfrac{3}{2} - \ln L_{i-1}\right]\end{aligned}\right\} \quad \text{for } (i-1)\text{th element} \tag{3.12}$$

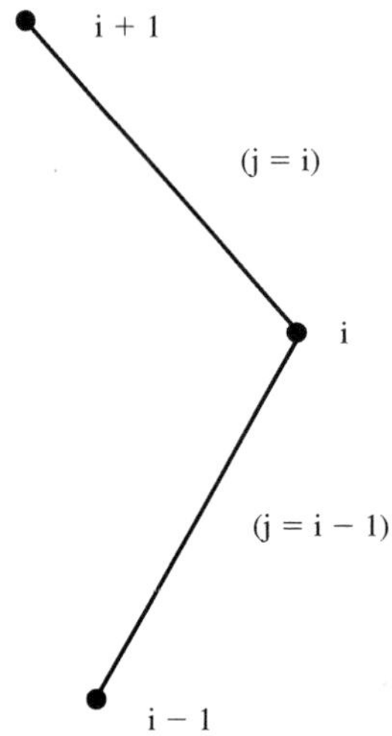

Figure 3.3. Linear singular boundary elements.

Evaluation of the jump term in equations (3.9) and (3.10)
It was mentioned earlier that the jump term C_i of equation (3.9) or (3.10) can be calculated in a simple way. Here we describe the procedure. We apply a uniform potential u over the whole boundary. Then, the v values are zero. Equation (2.25) becomes:

$$[H]\{u\} = \{0\}$$

Since $\{u\}$ is uniform, the sum of all the elements of $[H]$ in any row ought to be zero. Hence,

$$h_{ii} = -\sum_{\substack{j=1 \\ (j \neq i)}}^{N} h_{ij} \tag{3.13}$$

So, the diagonal of matrix $[H]$ can be computed indirectly without resorting to a geometric evaluation of the coefficient C_i. Once again, the application of appropriate boundary conditions to equation (2.25) will lead to equation (2.27).

3.3. Higher-order elements in 2-D

The formulation developed in the previous section can be systematically extended to quadratic and higher-order boundary elements. Higher-order boundary elements may be useful (a) to represent the geometry more accurately and (b) to obtain more accurate results for the solution variables. One can start from the discretized equation (3.1). The unknown quantities u and v will now be quadratic or higher-order functions of the coordinate η defined over the transformed domain $(-1, 1)$. For the quadratic boundary elements (Fig. 3.4), these functions are written as:

$$\left.\begin{aligned} u &= \phi_i(\eta)u_i \\ v &= \phi_i(\eta)v_i \end{aligned}\right\} \qquad (i = 1, 2, 3) \tag{3.14}$$

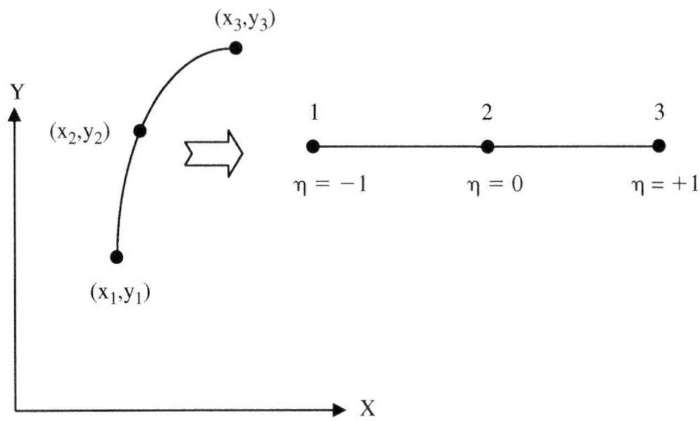

Figure 3.4. Quadratic boundary element and its transformation.

The quadratic shape functions ϕ_i are given by:

$$\left.\begin{aligned}\phi_1(\eta) &= \tfrac{1}{2}\eta(\eta-1)\\ \phi_2(\eta) &= \tfrac{1}{2}\eta(\eta+1)\\ \phi_3(\eta) &= (\eta-1)(\eta+1)\end{aligned}\right\} \tag{3.15}$$

In order to represent the differential boundary segment $\mathrm{d}\Gamma$ in terms of the coordinate η, the geometric coordinates x and y are also expressed using the shape functions of equation (3.15):

$$\left.\begin{aligned}x &= \phi_i(\eta)x_i\\ y &= \phi_i(\eta)y_i\end{aligned}\right\} \qquad (i = 1, 2, 3) \tag{3.16}$$

The differential boundary segment $\mathrm{d}\Gamma$ can now be written as:

$$\mathrm{d}\Gamma = |G|\mathrm{d}\eta \tag{3.17}$$

where, the Jocobian of transformation $|G|$, is given by:

$$|G| = \left[\left(\frac{\mathrm{d}x}{\mathrm{d}\eta}\right)^2 + \left(\frac{\mathrm{d}y}{\mathrm{d}\eta}\right)^2\right]^{\frac{1}{2}} \tag{3.18}$$

Thus, the integral involving the function u of the boundary element equation (3.1) becomes:

$$\int_{\Gamma_j} uv^*\mathrm{d}\Gamma = \int_{-1}^{1} [\phi_1 \quad \phi_2 \quad \phi_3]v^*|G|\,\mathrm{d}\eta \begin{Bmatrix}u_1\\ u_2\\ u_3\end{Bmatrix} = \left[\hat{h}_{i1} \quad \hat{h}_{i2} \quad \hat{h}_{i3}\right]\begin{Bmatrix}u_1\\ u_2\\ u_3\end{Bmatrix} \tag{3.19}$$

The integrals involving the function q can be dealt with in the same fashion.

Similarly, four-noded cubic boundary elements can be formulated using a cubic variation of the unknown functions u and v. The shape functions $\phi_i(\eta)$ in this case are given by:

$$\left.\begin{aligned} \phi_1(\eta) &= \tfrac{1}{16}(1-\eta)\left\{9(1+\eta^2)-10\right\} \\ \phi_2(\eta) &= \tfrac{1}{16}(1+\eta)\left\{9(1+\eta^2)-10\right\} \\ \phi_3(\eta) &= \tfrac{9}{16}(1-\eta^2)(1-3\eta) \\ \phi_4(\eta) &= \tfrac{9}{16}(1-\eta^2)(1+3\eta) \end{aligned}\right\} \tag{3.20}$$

3.4. Boundary elements in 3-D

In the BEM the three-dimensional problems can be solved using two-dimensional boundary elements. Here only the surface of the three-dimensional body needs to be discretized, thereby eliminating the need for the meshing of the volume. Both quadrilateral and triangular shaped elements are used. These elements are oriented in arbitrary three-dimensional space. We first need to perform a coordinate transformation from the global (x, y, z) system to the local (ξ, η, ζ) system. The coordinate axes ξ and η are tangential to the surface at the point under consideration and the ζ-axis is perpendicular to the surface. The derivatives of the function u with respect to the coordinates (x, y, z) can be expressed in terms of (ξ, η, ζ) from the relation:

$$\begin{Bmatrix} \dfrac{\partial u}{\partial \eta} \\ \dfrac{\partial u}{\partial \xi} \\ \dfrac{\partial u}{\partial \varsigma} \end{Bmatrix} = \boldsymbol{J} \begin{Bmatrix} \dfrac{\partial u}{\partial x} \\ \dfrac{\partial u}{\partial y} \\ \dfrac{\partial u}{\partial z} \end{Bmatrix} \tag{3.21}$$

where the Jacobian of the transformation $\boldsymbol{J}$ is given by:

$$\boldsymbol{J} = \begin{bmatrix} \dfrac{\partial x}{\partial \eta} & \dfrac{\partial y}{\partial \eta} & \dfrac{\partial z}{\partial \eta} \\ \dfrac{\partial x}{\partial \xi} & \dfrac{\partial y}{\partial \xi} & \dfrac{\partial z}{\partial \xi} \\ \dfrac{\partial x}{\partial \varsigma} & \dfrac{\partial y}{\partial \varsigma} & \dfrac{\partial z}{\partial \varsigma} \end{bmatrix} \tag{3.22}$$

The derivatives of the function u with respect to the coordinates (x, y, z) can be solved by inverting the relation (3.21):

$$\begin{Bmatrix} \dfrac{\partial u}{\partial x} \\ \dfrac{\partial u}{\partial y} \\ \dfrac{\partial u}{\partial z} \end{Bmatrix} = \boldsymbol{J}^{-1} \begin{Bmatrix} \dfrac{\partial u}{\partial \eta} \\ \dfrac{\partial u}{\partial \xi} \\ \dfrac{\partial u}{\partial \varsigma} \end{Bmatrix} \tag{3.23}$$

The typical boundary integral terms that need to be evaluated are:

$$\iint_S u^* v \mathrm{d}S \qquad \iint_S uv^* \mathrm{d}S \tag{3.24}$$

The differential area dS can be expressed in terms of the local coordinates (ξ, η, ζ). If $\boldsymbol{r}$ is the position vector of a point on the surface of the body, then dS is given by:

$$\mathrm{d}S = \left| \frac{\partial \boldsymbol{r}}{\partial \eta} \times \frac{\partial \boldsymbol{r}}{\partial \xi} \right| = |\boldsymbol{G}|\, \mathrm{d}\eta\, \mathrm{d}\xi \tag{3.25}$$

The vector cross product yields a vector that is perpendicular to the surface, i.e., along the ζ-axis and $|\boldsymbol{G}|$ is the magnitude of this normal. The derivatives $\partial \boldsymbol{r}/\partial \xi$ and $\partial \boldsymbol{r}/\partial \eta$ are written as:

$$\left.\begin{aligned} \frac{\partial \boldsymbol{r}}{\partial \eta} &= \left(\frac{\partial x}{\partial \eta}, \frac{\partial y}{\partial \eta}, \frac{\partial z}{\partial \eta} \right) \\ \frac{\partial \boldsymbol{r}}{\partial \xi} &= \left(\frac{\partial x}{\partial \xi}, \frac{\partial y}{\partial \xi}, \frac{\partial z}{\partial \xi} \right) \end{aligned}\right\} \tag{3.26}$$

The normal vector $\boldsymbol{n}$ can be computed from the vector cross product:

$$\boldsymbol{n} = \left| \frac{\partial \boldsymbol{r}}{\partial \eta} \times \frac{\partial \boldsymbol{r}}{\partial \xi} \right| = (g_1, g_2, g_3) \tag{3.27}$$

with the components:

$$\left.\begin{aligned} g_1 &= \frac{\partial x}{\partial \varsigma} = \frac{\partial y}{\partial \eta}\frac{\partial z}{\partial \xi} - \frac{\partial y}{\partial \xi}\frac{\partial z}{\partial \eta} \\ g_2 &= \frac{\partial y}{\partial \varsigma} = \frac{\partial z}{\partial \eta}\frac{\partial x}{\partial \xi} - \frac{\partial z}{\partial \xi}\frac{\partial x}{\partial \eta} \\ g_3 &= \frac{\partial z}{\partial \varsigma} = \frac{\partial x}{\partial \eta}\frac{\partial y}{\partial \xi} - \frac{\partial x}{\partial \xi}\frac{\partial y}{\partial \eta} \end{aligned}\right\} \tag{3.28}$$

The magnitude of $\boldsymbol{n}$ is then computed as:

$$|\boldsymbol{G}| = \sqrt{g_1^2 + g_2^2 + g_3^2} \tag{3.29}$$

The integrals of equation (3.24) are then converted to:

$$\iint_S u^* v |\boldsymbol{G}| \mathrm{d}\xi\, \mathrm{d}\eta \qquad \iint_S uv^* |\boldsymbol{G}|\, \mathrm{d}\xi\, \mathrm{d}\eta \tag{3.30}$$

The expression for a differential volume, dΩ can be obtained from a vector box product:

$$\mathrm{d}\Omega = \left| \frac{\partial \boldsymbol{r}}{\partial \eta} \times \frac{\partial \boldsymbol{r}}{\partial \xi} \cdot \frac{\partial \boldsymbol{r}}{\partial \varsigma} \right| \mathrm{d}\eta\, \mathrm{d}\xi\, \mathrm{d}\varsigma = |\boldsymbol{J}|\, \mathrm{d}\eta\, \mathrm{d}\xi\, \mathrm{d}\varsigma \tag{3.31}$$

which is utilized in evaluating the body load domain integral term of the equation (2.50):

$$\int bu^*\, \mathrm{d}\Omega = \int bu^* |\boldsymbol{J}|\, \mathrm{d}\eta\, \mathrm{d}\xi\, \mathrm{d}\varsigma \tag{3.32}$$

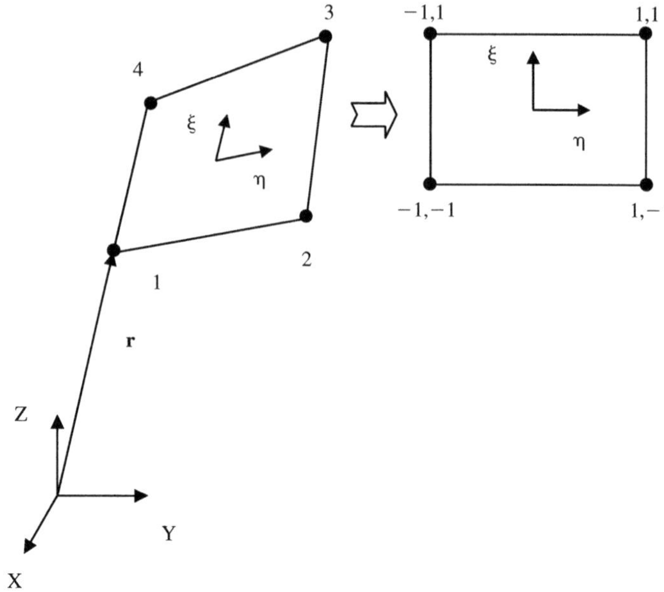

Figure 3.5. Four-noded quadrilateral boundary element and its transformation.

The above coordinate transformation allows one to develop three-dimensional boundary elements. The formulation for quadrilateral and triangular boundary elements is presented in the following sections.

3.4.1. Quadrilateral elements

The quadrilateral elements can be defined by four nodal points (Fig. 3.5) with bilinear variation of the unknown functions u and v. These functions as well as the coordinates (x, y, z) are expressed, much the same way as in equations (3.14) and (3.16), using shape functions $\phi(\xi,\eta)$ and nodal values of the unknown functions u_i (or v_i) or coordinates x_i (or y_i or z_i). To facilitate numerical integration, the shape functions are chosen such that the quadrilateral defined by the nodal coordinates $[(x_1, y_1), (x_2, y_2), (x_3, y_3), (x_4, y_4)]$ is mapped on to a domain in the (ξ, η) plane with the transformed nodal coordinates $[(-1,-1), (1,-1), (1, 1), (-1, 1)]$. The shape functions for the quadrilateral boundary elements are given by:

$$\left.\begin{aligned} \phi_1(\eta,\xi) &= \tfrac{1}{4}(1-\eta)(1-\xi) \\ \phi_2(\eta,\xi) &= \tfrac{1}{4}(1+\eta)(1-\xi) \\ \phi_3(\eta,\xi) &= \tfrac{1}{4}(1+\eta)(1+\xi) \\ \phi_4(\eta,\xi) &= \tfrac{1}{4}(1-\eta)(1+\xi) \end{aligned}\right\} \tag{3.33}$$

All the functions of the equations, (3.30) and (3.32) can now be evaluated at the Gaussian quadrature points.

Higher-order quadrilaterals can also be defined using more nodes and shape functions. For example, we can define an eight-noded serendipity quadrilateral element or a nine-noded Lagrangian quadrilateral element. The derivatives of the unknown

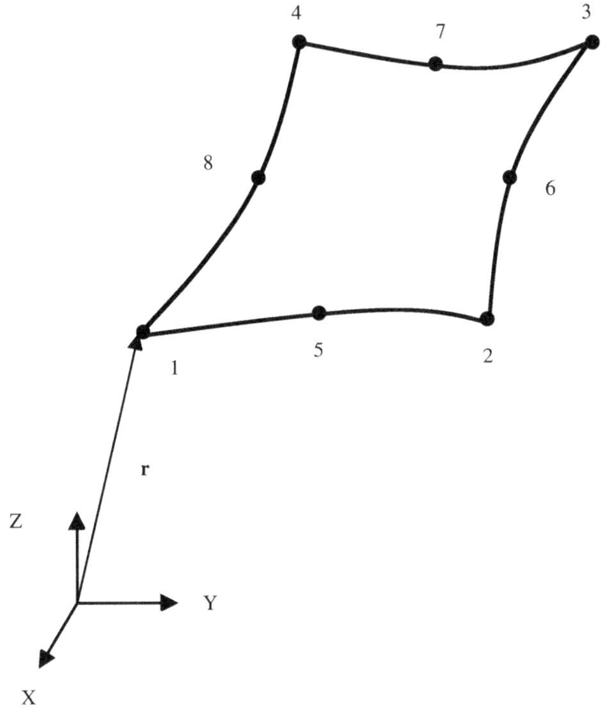

Figure 3.6. Eight-noded serendipity quadrilateral boundary element.

functions can also be used as nodal degrees of freedom. The shape functions for the eight-noded serendipity quadrilateral element (Fig. 3.6) are given below:

$$\left.\begin{aligned}
\phi_1(\eta,\xi) &= \tfrac{1}{4}(1-\eta)(1-\xi)(-\eta-\xi-1)\\
\phi_2(\eta,\xi) &= \tfrac{1}{4}(1+\eta)(1-\xi)(\eta-\xi-1)\\
\phi_3(\eta,\xi) &= \tfrac{1}{4}(1+\eta)(1+\xi)(\eta+\xi-1)\\
\phi_4(\eta,\xi) &= \tfrac{1}{4}(1-\eta)(1+\xi)(-\eta+\xi-1)\\
\phi_5(\eta,\xi) &= \tfrac{1}{2}(1-\eta^2)(1-\xi)\\
\phi_6(\eta,\xi) &= \tfrac{1}{2}(1-\xi^2)(1+\eta)\\
\phi_7(\eta,\xi) &= \tfrac{1}{2}(1-\eta^2)(1+\xi)\\
\phi_8(\eta,\xi) &= \tfrac{1}{2}(1-\xi^2)(1-\eta)
\end{aligned}\right\} \tag{3.34}$$

3.4.2. Triangular elements

A triangular boundary element can be defined by three nodal points (Fig. 3.7). A local oblique coordinate system (ξ, η) is defined at the nodal point 3. The local vectors $\boldsymbol{e}_1$ and $\boldsymbol{e}_2$, defined along ξ and η, respectively, can be expressed in terms of the global unit vector triad $(\boldsymbol{i}, \boldsymbol{j}, \boldsymbol{k})$:

$$\left.\begin{aligned}
\boldsymbol{e}_1 &= \tfrac{x_1-x_3}{L_{13}}\boldsymbol{i} + \tfrac{y_1-y_3}{L_{13}}\boldsymbol{j} + \tfrac{z_1-z_3}{L_{13}}\boldsymbol{k}\\
\boldsymbol{e}_2 &= \tfrac{x_2-x_3}{L_{23}}\boldsymbol{i} + \tfrac{y_2-y_3}{L_{23}}\boldsymbol{j} + \tfrac{z_2-z_3}{L_{23}}\boldsymbol{k}
\end{aligned}\right\} \tag{3.35}$$

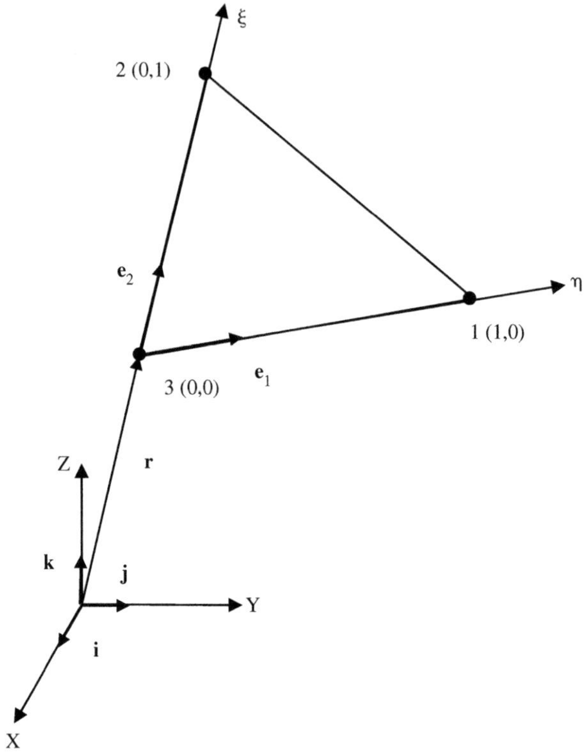

Figure 3.7. Three-noded triangular boundary element and its transformation.

L_{13} and L_{23} are the lengths of the sides 1–3 and 2–3 of the triangle. The position vector $\boldsymbol{r}$ on any point on the triangular region is given by:

$$\boldsymbol{r} = x\boldsymbol{i} + y\boldsymbol{j} + z\boldsymbol{k} = x_3\boldsymbol{i} + y_3\boldsymbol{j} + z_3\boldsymbol{k} + L_{13}\eta\boldsymbol{e}_1 + L_{23}\xi\boldsymbol{e}_2 \tag{3.36}$$

Substituting the expressions for the vectors $\boldsymbol{e}_1$ and $\boldsymbol{e}_2$ from equation (3.35) into equation (3.36), one obtains:

$$\boldsymbol{r} = x\boldsymbol{i} + y\boldsymbol{j} + z\boldsymbol{k} \tag{3.37}$$

with

$$\left.\begin{aligned} x &= \eta x_1 + \xi x_2 + (1 - \eta - \xi)x_3 \\ y &= \eta y_1 + \xi y_2 + (1 - \eta - \xi)y_3 \\ z &= \eta z_1 + \xi z_2 + (1 - \eta - \xi)z_3 \end{aligned}\right\} \tag{3.38}$$

From the above equation it can be seen that if we define a third coordinate ζ such that $\zeta = 1 - \xi - \eta$, then we can write:

$$\left.\begin{aligned} x &= \eta x_1 + \xi x_2 + \varsigma x_3 \\ y &= \eta y_1 + \xi y_2 + \varsigma y_3 \\ z &= \eta z_1 + \xi z_2 + \varsigma z_3 \end{aligned}\right\} \tag{3.39}$$

which can be written in matrix form:

$$\begin{Bmatrix} x \\ y \\ z \end{Bmatrix} = \begin{bmatrix} x_1 & x_2 & x_3 \\ y_1 & y_2 & y_3 \\ z_1 & z_2 & z_3 \end{bmatrix} \begin{Bmatrix} \eta \\ \xi \\ \varsigma \end{Bmatrix} \tag{3.40}$$

Since ζ is not an independent coordinate, we can only invert the first two equations:

$$\left.\begin{aligned} \eta &= \tfrac{1}{2A}\left[2A_1 + \beta_1 x + \alpha_1 y\right] \\ \xi &= \tfrac{1}{2A}\left[2A_2 + \beta_2 x + \alpha_2 y\right] \end{aligned}\right\} \tag{3.41}$$

The coordinate ζ is obtained from the relation $\zeta = 1 - \xi - \eta$, yielding

$$\varsigma = \frac{1}{2A}\left[2A_3 + \beta_3 x + \alpha_3 y\right] \tag{3.42}$$

The expressions for α_i, β_i and A_i can be obtained from the following recursive relations:

$$\left.\begin{aligned} \alpha_i &= x_k - x_j \\ \beta_i &= y_j - y_k \\ 2A_i &= x_j y_k - x_k y_j \end{aligned}\right\} \quad (i = 1, 2, 3; \quad j = 2, 3, 1; \quad k = 3, 1, 2) \tag{3.43}$$

The area of the triangle A is computed as:

$$A = \frac{1}{2}(\beta_1 \alpha_2 - \beta_2 \alpha_1) \tag{3.44}$$

This area is the projection of the triangular element on to the (x, y) plane. It is obvious from equations (3.39) that the interpolation functions in this case are simply the local coordinates ($\phi_1 = \xi$, $\phi_2 = \eta$ and $\phi_3 = \zeta$). Thus,

$$u = \phi_i u_i = \eta u_1 + \xi u_2 + \varsigma u_3 \tag{3.45}$$

Because of the use of local coordinates (ξ, η, and ζ) the necessary integrals can be performed here in closed form:

$$\iint \eta^i x^j \varsigma^k \, \mathrm{d}A = \frac{i!\, j!\, k!}{(i + j + k)!}(2A) \tag{3.46}$$

The shape functions for the six-noded second-order triangular elements (Fig. 3.8) are given as follows:

$$\left.\begin{aligned} \phi_1 &= \eta(2\eta - 1) \\ \phi_2 &= \xi(2\xi - 1) \\ \phi_3 &= \varsigma(2\varsigma - 1) \\ \phi_4 &= 4\eta\xi \\ \phi_5 &= 4\xi\varsigma \\ \phi_6 &= 4\varsigma\eta \end{aligned}\right\} \tag{3.47}$$

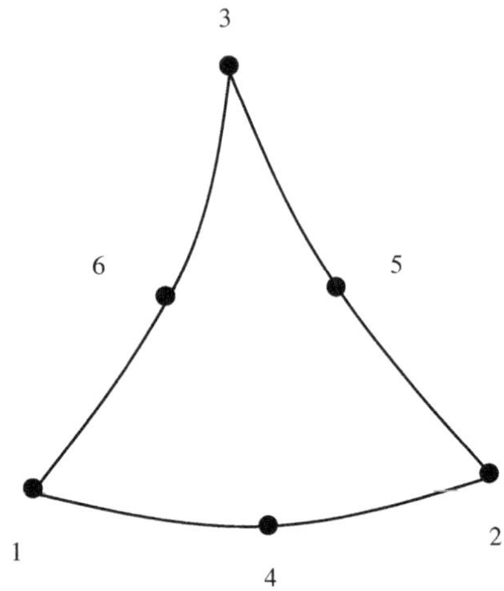

Figure 3.8. Six-noded serendipity triangular boundary element.

3.5. Examples

A number of example problems are presented in this section illustrating the use of lower and higher order boundary elements in 2-D and 3-D. For simplicity, the theoretical developments of this and last chapter were confined to potential problems only. The formulations presented so far are however expandable to elasticity or acoustics. Although we are postponing the development of detailed boundary element formulations for elasticity and acoustics till later chapters, we present two elasticity problems in this section. These elastostatic stress analysis problems are designed to demonstrate the use and performance of higher order boundary elements.

Example 3.1: Flow of perfect fluid around aerofoils
In this example we will study the flow of a perfect fluid past an aerofoil designated as NACA 0018. The flow can be indirectly described by a stream function u, which is related to the velocities in the following manner:

$$v_x = \frac{\partial u}{\partial x}, \quad v_y = -\frac{\partial u}{\partial x} \tag{3.48}$$

The far field velocities away from the aerofoil are given as:

$$v_x = V, \quad v_y = 0 \tag{3.49}$$

The stream function u can be separated into two components:

$$u = u_1 + u_2 \tag{3.50}$$

where u_1 is the free flow stream function in the absence of the aerofoil and u_2 is the stream function for perturbed flow. Thus, $u_1 = Vy$. If we consider the total stream function $u = 0$ on the aerofoil surface Γ, then

$$u_2 = -u_1 = -Vy \quad \text{on } \Gamma \tag{3.51}$$

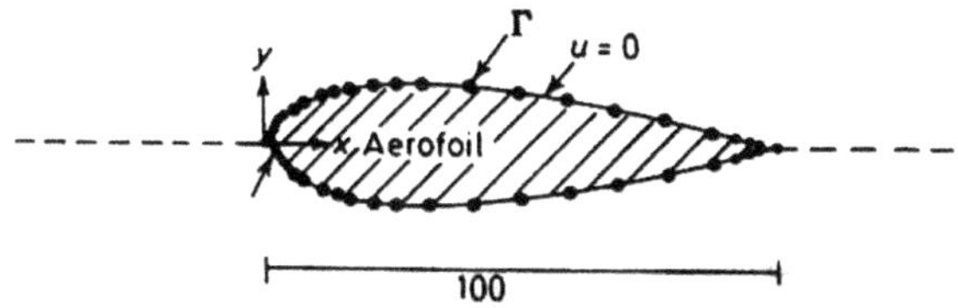

Figure 3.9. Boundary element discretization of NACA 0018 aerofoil.

Table 3.1. Tangential velocity of flow around NACA 0018 aerofoil.

Solution for v/V ($v =$ tangential velocity)		
x-coordinate	Analytical solution	Boundary element solution
0.0	0.000	0.000
1.25	0.926	0.931
2.5	1.103	1.093
5.0	1.228	1.212
7.5	1.267	1.253
10.0	1.276	1.265
15.0	1.278	1.273
20.0	1.275	1.269
25.0	1.262	1.254
30.0	1.247	1.236
40.0	1.205	1.198
50.0	1.157	1.158
60.0	1.116	1.118
70.0	1.074	1.079
80.0	1.025	1.036
90.0	0.966	0.975
95.0	0.914	0.941
100.0	0.000	0.000

Due to symmetry we only need to discretize one-half of the aerofoil (Fig. 3.9). Table 3.1 shows the tangential velocity, normalized to V, as a function of x-coordinate solved using linear boundary elements. The solution produced by NACA is also shown in the table [69].

Example 3.2: Heat conduction in an infinite medium
Consider a spherical cavity of unit radius placed in an infinite conducting medium [71]. A constant radial flux of 10 J/(m^{2}s) is prescribed over the surface of the cavity. The temperature distribution in the infinite medium can be computed using 3-D boundary element discretization. Unlike the FEM, here only the surface of the cavity is required to be discretized. Due to symmetry, only one-eighth of the cavity surface is modeled using flat triangular elements with constant potential and normal derivative. Unlike finite methods, the use of constant field variables and its normal derivatives on the element segment is commonplace in the boundary element literature. The inter-element discontinuities in solution variables are mitigated by the fact that BEM is an integral equation technique and tends to average out errors arising out of discontinuities.

Table 3.2. Temperature distribution in an infinite medium from a spherical cavity.

R	BEM ($N=7$)	BEM ($N=16$)	Exact
1.0	9.676	9.727	10.000
1.5	6.505	6.569	6.667
2.0	4.899	4.922	5.000
3.0	3.274	3.281	3.333
6.0	1.639	1.640	1.667
10.0	0.983	0.984	1.000
100.0	0.098	0.098	0.100
1000.0	0.010	0.010	0.010

($N=$ number of boundary element segments).

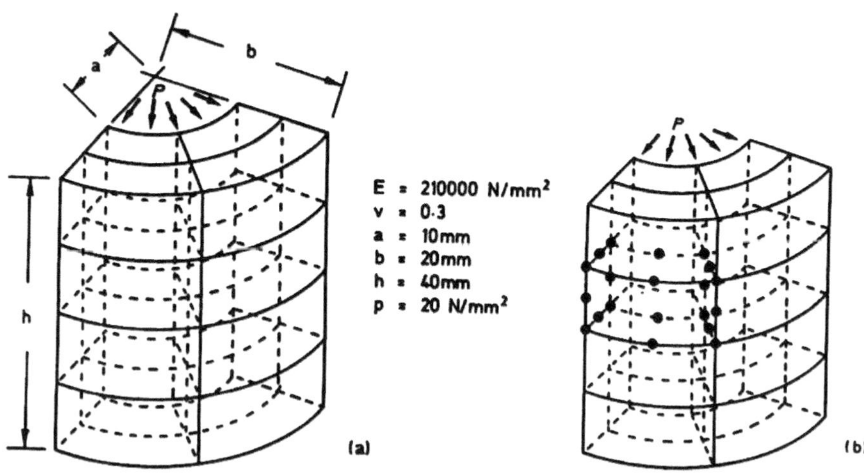

Figure 3.10. (a) Boundary element surface mesh and (b) finite element 3-D mesh of thick cylinder.

Results, produced using constant elements, are presented in Table 3.2, where averaged temperatures on the cavity surface and temperature distribution inside the infinite medium are shown. The exact solution for this problem is given by $u=10/R$, where R is the radial distance ($R \geq 1$). The exact solutions are also given in the table. The boundary element (BE) solution would converge at a faster rate to the exact solution if higher order boundary elements were employed.

Example 3.3: Internally pressurized thick cylinder

A thick cylinder of length $h=40$ mm, having inner and outer diameters $a=10$ mm and $b=20$ mm respectively, is subjected to internal pressure $p=20\,\text{N/mm}^2$. The Young's modulus $E=210000\,\text{N/mm}^2$ and Poisson's ratio $\upsilon=0.3$ for the cylinder material. A 90° sector is modeled using (a) linear boundary element, (b) quadratic boundary element and (c) finite element [63]. The finite element and boundary element meshes are shown in Figure 3.10. A 20-noded serendipity element is used for finite element discretization.

Table 3.3. Displacements and stresses in internally pressurized thick cylinder.

Function	Radius	Exact solution	Finite element solution	Boundary element solution	
				Linear BE	Quadratic BE
Radial displacement	10.0	1.904	1.905	1.818	1.905
	12.5	1.602	1.600	1.568	1.600
	15.0	1.415	1.414	1.319	1.414
	17.5	1.293	1.292	1.234	1.291
	20.0	1.212	1.212	1.150	1.211
Radial stress	10.0	−20.0	−17.4	−13.8	−18.5
	12.5	−10.4	−11.6	−13.2	−11.3
	15.0	−5.2	−3.5	−1.8	−4.1
	17.5	−2.0	−2.4	−1.2	−2.2
	20.0	0.0	0.7	1.1	0.4
Hoop stress	10.0	33.3	34.4	27.1	33.5
	12.5	23.7	23.3	20.1	23.7
	15.0	18.5	19.2	15.2	18.5
	17.5	15.4	15.2	14.3	15.4
	20.0	13.3	13.5	12.1	13.3

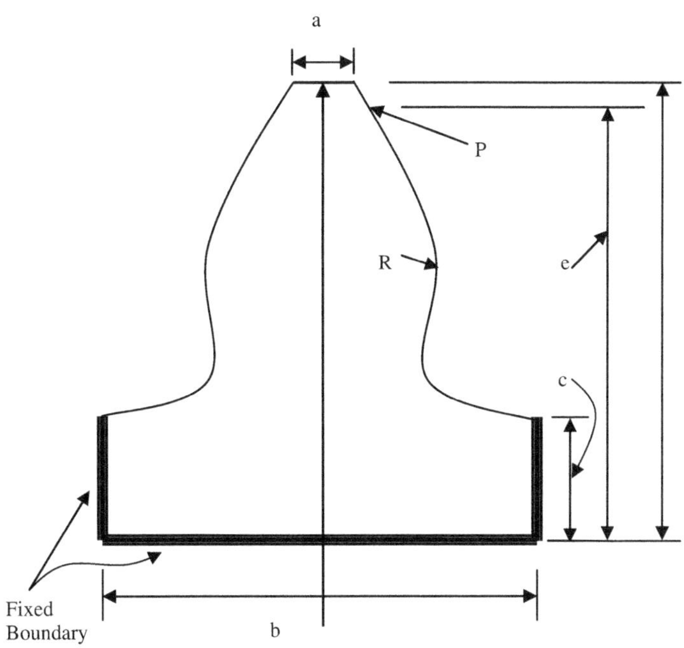

Figure 3.11. Schematic of the gear tooth. $a = 2.00$ mm, $b = 16.00$ mm, $c = 5.70$ mm, $d = 18.25$ mm, $e = 17.25$ mm, R (radius) $= 31.25$ mm.

A four-noded surface element and an eight-noded serendipity surface element are used for the linear and quadratic boundary element models respectively. In the case of boundary element solution, the numerical integration points were concentrated near singularities to achieve better accuracy [63].

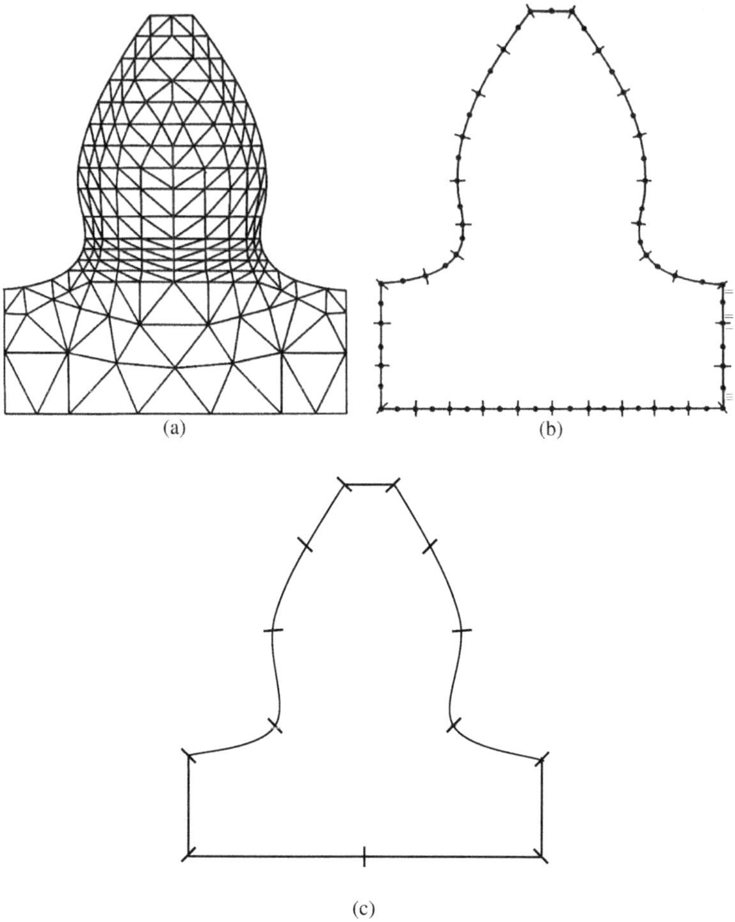

Figure 3.12. (a) Finite element discretization of the gear tooth. (b) Linear and quadratic BE discretizations of the gear tooth. (c) Cubic BE discretizations of the gear tooth.

The results are shown in Table 3.3 where the exact solutions are also included. Whereas linear BE solutions are poorer than those of FE (finite element), the quadratic BE solutions are seen to be in good agreement with the exact solutions.

Example 3.4: Stress in a gear tooth
A gear tooth with dimensions, load and boundary conditions is shown in Figure 3.11. The load P acts normal to the gear tooth surface at the point shown in the figure and equals 400 N/mm. This problem was solved by Lachat [60] and later by Brebbia [63]. Assuming plane strain condition, the tooth was analyzed using (a) 291 six-noded isoparametric triangular finite elements with 630 nodal points (Fig. 3.12a), (b) 33 linear BE segments, (c) 33 quadratic BE segments (Fig. 3.12b), (d) 33 cubic BE segments (Fig. 3.12b) and finally (e) 13 cubic BE segments (Fig. 3.12c). None of these idealizations produced reasonable stresses at the point of load application. The principal stress at the other points on the gear tooth surface is shown in Figure 3.13 for FE, quadratic BE and cubic BE discretizations.

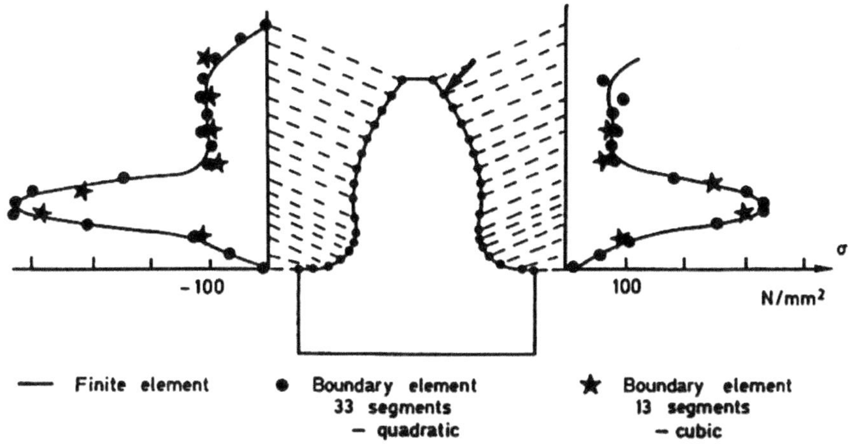

Figure 3.13. Principal stresses at the surface of the gear tooth.

Chapter 4

Anisotropy, Axisymmetry and Zoning

4.1. Introduction

Unlike the finite elements, the boundary element method (BEM) requires relatively more analytical work to formulate the numerical equations in the case of axisymmetric bodies and when the material anisotropy needs to be considered.

When the solution domain contains inhomogeneity, it can be handled in BEM by partitioning the domain into zones having homogeneous properties. As an added benefit, zoning produces banded system matrix thereby making the problem suitable for banded matrix solvers. We will see in later chapters that zoning helps improve the solution accuracy, especially when dealing with acoustic eigenvalue analysis of chunky-type enclosures.

4.2. Anisotropic materials

At times engineering materials and media cannot be adequately described using isotropic material properties. The materials may respond differently in orthogonal directions (orthotropic behavior) or in all directions (anisotropic behavior). In this section, we shall present necessary ingredients which will allow us to formulate boundary integral equations in these cases.

Orthotropic materials

Consider the domain shown in Figure 4.1 with x and y being the directions of orthotropy. The Laplace's equation in this case is given by:

$$k_x \frac{\partial^2 u}{\partial x^2} + k_y \frac{\partial^2 u}{\partial y^2} = 0 \qquad \text{(2-D)} \tag{4.1}$$

$$k_x \frac{\partial^2 u}{\partial x^2} + k_y \frac{\partial^2 u}{\partial y^2} + k_z \frac{\partial^2 u}{\partial z^2} = 0 \qquad \text{(3-D)} \tag{4.2}$$

where k_x, k_y and k_z are the material properties in the x, y and z directions, respectively. The fundamental solutions to these equations are:

$$u^* = \frac{1}{2\pi\sqrt{k_x k_y}} \ln\left(\frac{1}{r}\right) \qquad \text{(2-D)} \tag{4.3}$$

$$u^* = \frac{1}{4\pi r\sqrt{k_x k_y k_z}} \qquad \text{(3-D)} \tag{4.4}$$

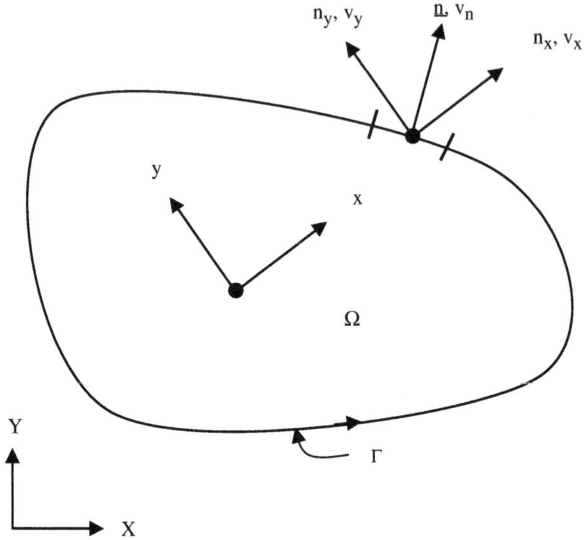

Figure 4.1. A Two-dimensional orthotropic medium (x & y are directions of orthotropy).

respectively, where:

$$r = \left[\frac{1}{k_x}\{x(p) - x(q)\}^2 + \frac{1}{k_y}\{y(p) - y(q)\}^2\right]^{\frac{1}{2}} \quad \text{(2-D)} \tag{4.5}$$

$$r = \left[\frac{1}{k_x}\{x(p) - x(q)\}^2 + \frac{1}{k_y}\{y(p) - y(q)\}^2 + \frac{1}{k_z}\{z(p) - z(q)\}^2\right]^{\frac{1}{2}} \quad \text{(3-D)} \tag{4.6}$$

It would also be necessary to define boundary fluxes v and v^*. To this end, we apply the divergence theorem to the left-hand side of equation (4.1):

$$\iint_\Omega \left(k_x \frac{\partial^2 u}{\partial x^2} + k_y \frac{\partial u}{\partial y}\right) \mathrm{d}\Omega = \int_\Gamma \left(k_x \frac{\partial u}{\partial x} n_x + k_y \frac{\partial u}{\partial y} n_y\right) \mathrm{d}\Gamma \quad \text{(2-D)} \tag{4.7}$$

where n_x and n_y are direction cosines and the quantity between the brackets on the right-hand side is the normal boundary flux:

$$v = k_x \frac{\partial u}{\partial x} n_x + k_y \frac{\partial u}{\partial y} n_y \qquad \text{(2-D)} \tag{4.8}$$

We can analogously define v^* as:

$$v^* = k_x \frac{\partial u^*}{\partial x} n_x + k_y \frac{\partial u^*}{\partial y} n_y \qquad \text{(2-D)} \tag{4.9}$$

The corresponding quantities in three-dimensions can easily be defined. Using the governing equation (4.1) or (4.2) and given boundary conditions, we can formulate the boundary integral equation in the same fashion as in the isotropic case and arrive at a formulation similar to equation (2.19).

Anisotropic materials
If the medium is anisotropic with the material property coefficients given by k_{ij}, the governing differential equation in two dimensions can be written as:

$$k_{xx}\frac{\partial^2 u}{\partial x^2} + 2k_{xy}\frac{\partial^2 u}{\partial x \partial y} + k_{yy}\frac{\partial^2 u}{\partial y^2} = 0 \tag{4.10}$$

The fundamental solution for this equation is:

$$u^* = \frac{1}{\sqrt{|k_{ij}|}} \ln\left(\frac{1}{r}\right) \tag{4.11}$$

where $|k_{ij}|$ is the determinant of the material property coefficient matrix and "r" is given by:

$$r = \left[\frac{1}{k_{xx}}\{x(p) - x(q)\}^2 + \frac{2}{k_{xy}}\{x(p) - x(q)\}\{y(p) - y(q)\} + \frac{1}{k_{yy}}\{y(p) - y(q)\}^2\right]^{\frac{1}{2}} \tag{4.12}$$

The quantities v and v^* are respectively given by:

$$v = \left(k_{xx}\frac{\partial u}{\partial x} + k_{xy}\frac{\partial u}{\partial y}\right) n_x + \left(k_{xy}\frac{\partial u}{\partial x} + k_{yy}\frac{\partial u}{\partial y}\right) n_y \tag{4.13}$$

$$v^* = \left(k_{xx}\frac{\partial u^*}{\partial x} + k_{xy}\frac{\partial u^*}{\partial y}\right) n_x + \left(k_{xy}\frac{\partial u^*}{\partial x} + k_{yy}\frac{\partial u^*}{\partial y}\right) n_y \tag{4.14}$$

The problem can be formulated as in the previous case.

4.3. Axisymmetric problems

Three-dimensional boundary value problems having axial symmetry in geometry, loading and boundary conditions can be solved in two dimensions, thereby saving significant efforts. A body with symmetry about the Z-axis is shown in Figure 4.2. In the BEM we only need to discretize the contour lines Γ_b. Thus, three-dimensional axisymmetric problems can be solved using line elements. To this end, the boundary integral equation (2.19) is written in cylindrical polar coordinate system (R, Θ, Z) with the fundamental solution u^* of equation (2.9) also expressed in the same coordinate system. Taking advantage of axial symmetry, the dependence on Θ is then integrated out. The final equation is given in the (R–Z) plane:

$$C(P)u(P) + \int_{\Gamma_b} u(Q)v^*(P,Q)r(Q)\,\mathrm{d}\Gamma(Q) = \int_{\Gamma_b} v(Q)u^*(P,Q)r(Q)\,\mathrm{d}\Gamma(Q) \tag{4.15}$$

in which the fundamental solution u^* and its normal derivative v^* are given by:

$$u^*(P,Q) = \frac{4K(\kappa)}{\left[\{r(P) + r(Q)\}^2 + \{z(P) - z(Q)\}^2\right]^{\frac{1}{2}}} \tag{4.16}$$

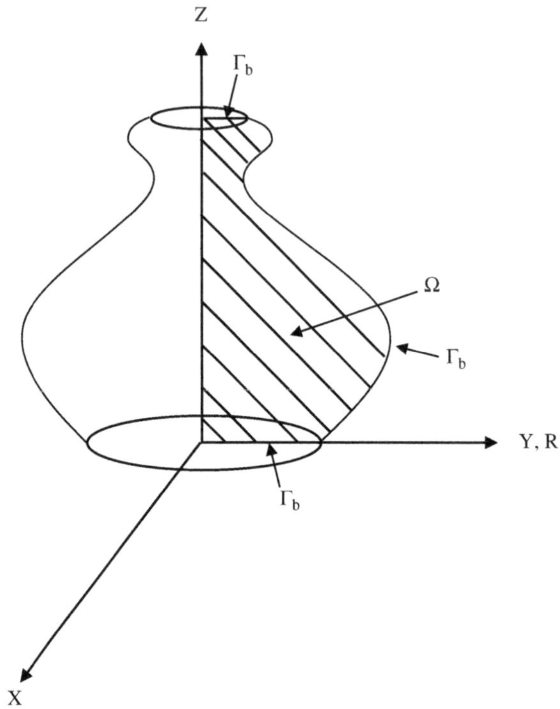

Figure 4.2. Axisymmetric body. Ω is the generating area and Γ_b is the boundary contour.

$$v^*(P,Q) = \frac{4}{\left[\{r(P)+r(Q)\}^2 + \{z(P)-z(Q)\}^2\right]^{\frac{1}{2}}}$$
$$\times \left[\frac{1}{2r(Q)} \left\{ \frac{r^2(P) - r^2(Q) + \{z(P)-z(Q)\}^2}{\{r(P)-r(Q)\}^2 + \{z(P)-z(Q)\}^2} E(\kappa) - K(\kappa) \right\} n_r(Q) \right.$$
$$\left. + \frac{z(P)-z(Q)}{\{r(P)-r(Q)\}^2 + \{z(P)-z(Q)\}^2} E(\kappa) n_z(Q) \right] \tag{4.17}$$

$K(\kappa)$ and $E(\kappa)$ are the complete elliptic integral of the first and second kinds respectively. The argument κ is given by:

$$\kappa = \frac{4r(P)r(Q)}{\{r(P)+r(Q)\}^2 + \{z(P)-z(Q)\}^2} \qquad (0 \le \kappa \le 1) \tag{4.18}$$

In equation (4.17), $n_r(Q)$ and $n_z(Q)$ are the R- and Z-components of the outward normal at the boundary field point Q. Note that the fundamental solution [eqn. (4.16)] in the axisymmetric case is a function of the distances of the source and field points P and Q from the axis of revolution Z. Recall that the fundamental solutions for the two- and three-dimensional cases [eqns. (2.8) and (2.9)] are given simply as a function of distance between the source and field points $r(P,Q)$. The solution of the equation (4.15) can be approached in the same fashion as was done in the case of the boundary element equation (2.18). The boundary contour Γ_b of the axisymmetric body of Figure 4.2 can be divided into a number of line segments. Numerical integration is then performed

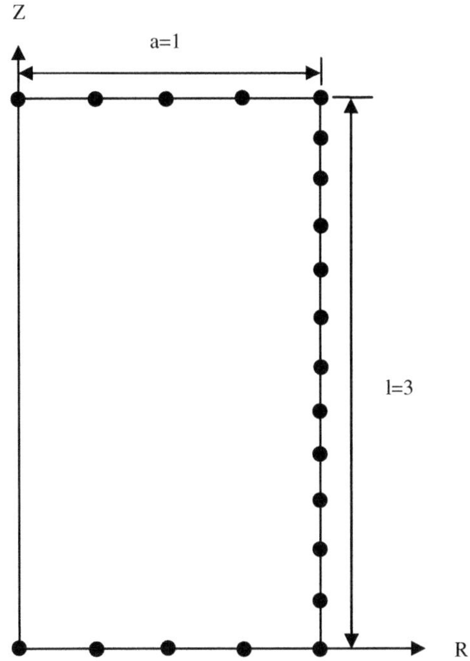

Figure 4.3. BE discretization of the axisymmetric solid cylinder.

on each line segment and contributions are assembled into a system of equation, which is of the form of equation (2.25).

When the source point P does not lie on or is close to an element segment, numerical integrals can be performed using a standard Gaussian quadrature. The values of the complete elliptic integrals can be incorporated into the computer program in tabular form as a function of their argument κ. For a computed value of the argument κ at a Gaussian point, the values of the complete integrals can be looked up from this table. However, the elliptic integrals can also be approximated by polynomial expressions [163]. For the evaluation of the integrals on the singular elements, the fundamental solution u^* and its normal derivative v^* are expressed in terms of Legendre functions of the second kind [71]. The integration can then be performed analytically by expanding these Legendre functions [164]. Furthermore, when $r(P)$ is small, i.e., when the element is located near the axis of revolution, the integration must be performed with care [71, 165].

Example 4.1: Heat conduction in a solid axisymmetric cylinder
Consider a solid axisymmetric cylinder with $0 \le R < a$ and $0 < Z < l$ [71]. Specified boundary conditions are given as: $u = 0$ at $Z = l$, $u = 1$ at $Z = 0$ and $v + hu = 0$ at $R = a$. The surface at $R = a$ has a convection boundary condition where h is the heat transfer coefficient. For this example, we assume $a = 1$, $l = 3$ and $h = 0.1$. In order to solve this problem by the BEM, only the surfaces at $Z = 0$, $Z = l$ and $R = a$ need to be discretized. A total of 20 equal-length linear boundary elements are used for this purpose (Fig. 4.3). Results are presented in Table 4.1 where exact solutions are also shown.

Table 4.1. Temperature distribution in a solid axisymmetric cylinder.

Z	BE solution ($R=0.25$)	Exact solution ($R=0.25$)	BE solution ($R=1.00$)	Exact solution ($R=1.00$)
0.5	0.781	0.781	0.751	0.751
1.0	0.585	0.585	0.560	0.560
1.5	0.416	0.416	0.397	0.397
2.0	0.267	0.267	0.254	0.254
2.5	0.130	0.130	0.124	0.124

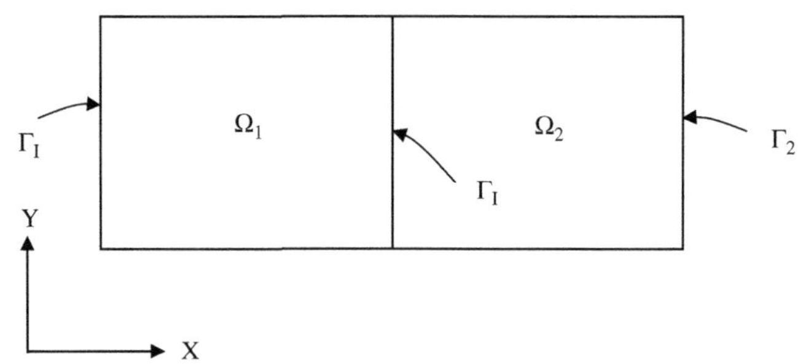

Figure 4.4. A domain is divided into two sub-regions (zones).

4.4. Inhomogeneous regions and zoning

In many practical engineering applications, the problem may only be piecewise homogeneous or the domain may have an irregular shape having one of the dimensions much larger than others. In these cases, the domain is divided into regular homogeneous subregions. The boundary element equations are derived independently for each subregion. These equations can be assembled into a single set of equations for the whole body using the compatibility and equilibrium conditions on the interface between the subregions.

Let us consider a Ω domain, which is divided into two different subregions Ω_1 and Ω_2 as shown in Figure 4.4. For the subregion 1, the boundary element equations can be written as:

$$[H^1 \quad H_I^1]\begin{Bmatrix} u^1 \\ u_I^1 \end{Bmatrix} = [G^1 \quad G_I^1]\begin{Bmatrix} v^1 \\ v_I^1 \end{Bmatrix} \tag{4.19}$$

The equations for the subregion 2 can similarly be written as:

$$[H^2 \quad H_I^2]\begin{Bmatrix} u^2 \\ u_I^2 \end{Bmatrix} = [G^2 \quad G_I^2]\begin{Bmatrix} v^2 \\ v_I^2 \end{Bmatrix} \tag{4.20}$$

where:

u^1, v^1 = nodal potentials and fluxes on Γ_1.
u^2, v^2 = nodal potentials and fluxes on Γ_2.
u_I^1, v_I^1 = nodal potentials and fluxes on Γ_I with respect to the subregion Ω_1.
u_I^2, v_I^2 = nodal potentials and fluxes on Γ_I with respect to the subregion Ω_2.

The flux $v = k\partial u/\partial n$. As mentioned earlier, the compatibility and equilibrium conditions must be satisfied at the interface boundary Γ_I:

(a) Compatibility condition: $u_I^1 = u_I^2 = u_I$ (4.21)

(b) Equilibrium condition: $v_I^1 = -v_I^2 = v_I$ (4.22)

where u_I and v_I denote the nodal potentials and fluxes on the interface boundary Γ_I. Utilizing these conditions, equations (4.19) and (4.20) can be assembled into a single system:

$$\begin{bmatrix} H^1 & H_I^1 & 0 \\ 0 & H_I^2 & H^2 \end{bmatrix} \begin{Bmatrix} u^1 \\ u_I \\ u^2 \end{Bmatrix} = \begin{bmatrix} G^1 & G_I^1 & 0 \\ 0 & -G_I^2 & G^2 \end{bmatrix} \begin{Bmatrix} v^1 \\ v_I \\ v^2 \end{Bmatrix} \tag{4.23}$$

Note that this system of equations can be written exactly as equation (2.25). One important feature of the system of equations (4.23) is that it is banded. If the region Ω were progressively divided into more and more subregions, this system would become more and more banded. The system matrix (4.23) remains banded even after application of the boundary conditions. The quantities u_I and v_I on the interface boundary are always unknowns for the problem.

Chapter 5

Time-Harmonic Analysis in Acoustics and Elasticity

5.1. Introduction

This chapter will briefly present the boundary element formulation of the time-harmonic problems in acoustics and elasticity. The behavior of a dynamic system to a transient excitation where the forcing function varies sinusoidally is an important consideration in both structural and acoustic designs. We know that the response to such an excitation will also be sinusoidal. As a result, the given problem is simplified since the "time" as a variable is replaced by the frequency of oscillation, i.e., the problem is treated in the frequency domain. We will first develop the boundary element formulation of the acoustics time-harmonic analysis as it involves only scalar equations.

5.2. Acoustics

The acoustic wave equation governing the acoustic pressure $\psi(\boldsymbol{x}, t)$ in a compressible fluid medium is given by

$$\nabla^2 \Psi = \frac{1}{c^2} \frac{\partial^2 \Psi}{\partial t^2} \tag{5.1}$$

where c is the speed of sound and t is the time. For the time harmonic oscillations of the sound pressure wave, we let $\Psi = u e^{j\omega t}$. u is the amplitude of the pressure, ω is the circular frequency of oscillation and $\mathrm{j} = \sqrt{-1}$. Substituting for Ψ in equation (5.1), the Helmholtz equation governing the amplitude of pressure oscillations is obtained

$$\nabla^2 u + k^2 u = 0 \tag{5.2}$$

k is the wave number given by ω / c. If we were employing the finite element method (FEM), we would straightaway employ the coefficient matrices, "stiffness" and "mass", derived from the initial weighted residual discretization of the wave equation (5.1). In boundary elements, however, the weighting function employed is unique to a given differential equation. As a result, we need to apply the weighted residual formulation to the Helmholtz equation (5.2) in order to generate boundary element (BE) coefficient matrices.

The boundary element formulation of this Helmholtz equation proceeds in the same manner as in the case of the Laplace's equation (2.1). The weighted residual statement for the Helmholtz equation corresponding to equation (2.2) can be written as:

$$\int_{\Omega} \left(\nabla^2 u + k^2 u\right) u^* \, \mathrm{d}\Omega = \int_{\Gamma_v} (v - v_b)\, u^* \, \mathrm{d}\Gamma - \int_{\Gamma_u} (u - u_b)\, v^* \, \mathrm{d}\Gamma \tag{5.3}$$

After applying the integral transformation of equation (5.3) twice, we obtain:

$$\int_{\Omega} u \left(\nabla^2 u^* + k^2 u^*\right) \mathrm{d}\Omega = -\int_{\Gamma_v} v_b u^* \, \mathrm{d}\Gamma - \int_{\Gamma_u} v u^* \, \mathrm{d}\Gamma + \int_{\Gamma_v} u v^* \, \mathrm{d}\Gamma + \int_{\Gamma_u} u_b v^* \, \mathrm{d}\Gamma \tag{5.4}$$

By definition, the fundamental solution u^* for the Helmholtz equation can be found by applying a unit singular source at the point "p". Thus, u^* should satisfy:

$$\nabla^2 u^* + k^2 u^* + \delta_i = 0 \tag{5.5}$$

Substituting this back in equation (5.4), dropping, for the time being, the distinction between the given boundary conditions (u_b and v_b) and the unknown quantities (u and v) and recognizing the fact that $\Gamma_u + \Gamma_v = \Gamma$, we arrive at the following boundary integral statement:

$$u_p + \int_{\Gamma} u v^* \mathrm{d}\Gamma = \int_{\Gamma} u^* v \, \mathrm{d}\Gamma \tag{5.6}$$

By taking the point "p" to the boundary, we obtain the final boundary element equation:

$$C_P u_P + \int_{\Gamma} u v^* \mathrm{d}\Gamma = \int_{\Gamma} u^* v \, \mathrm{d}\Gamma \tag{5.7}$$

The fundamental solutions [those which satisfy the differential eqn. (5.5)] and their normal derivatives u^* and v^*, respectively, of the Helmholtz equation are given by:

$$u^* = \frac{\mathrm{j}}{4} H_0^1(kr) = \frac{\mathrm{j}}{4}[J_0(kr) + \mathrm{j}Y_0(kr)] \qquad \text{(2-D)} \tag{5.8}$$

$$v^* = -\frac{\mathrm{j}k}{4} H_1^1(kr) \frac{\partial r}{\partial n} = -\frac{\mathrm{j}k}{4}\left[J_1(kr) + \mathrm{j}Y_1(kr)\right] \frac{\partial r}{\partial n} \qquad \text{(2-D)} \tag{5.9}$$

$$u^* = \frac{1}{4\pi r} \mathrm{e}^{-\mathrm{j}kr} \qquad \text{(3-D)} \tag{5.10}$$

$$v^* = \frac{1}{4\pi r}\left(\frac{1}{r} - \mathrm{j}k\right) \mathrm{e}^{-\mathrm{j}kr} \frac{\partial r}{\partial n} \qquad \text{(3-D)} \tag{5.11}$$

These fundamental solutions satisfy the Sommerfeld radiation condition at infinity, which is expressed as:

$$\sqrt{r}\left(\frac{\partial u}{\partial r} + \mathrm{j}ku\right) \to 0 \quad \text{as } r \to \infty \qquad \text{(2-D)} \tag{5.12}$$

$$r\left(\frac{\partial u}{\partial r} + \mathrm{j}ku\right) \to 0 \quad \text{as } r \to \infty \qquad \text{(3-D)} \tag{5.13}$$

We first substitute the fundamental solution u^* and its normal derivative v^* from equations (5.8) through (5.11) into the boundary element equation (5.7). The boundary is then discretized into boundary element segments. Introducing appropriate polynomial shape functions, as outlined in Chapter 2, integration can be performed over all of the segments on the boundary leading to the boundary element matrix equation:

$$[H]\{u\} = [G]\{v\} \tag{5.14}$$

The matrices $[H]$ and $[G]$ are complex. Note that the fundamental solutions contain the frequency parameter "k" as arguments of transcendental functions sin(), cos() and ln(). As a result, "k" cannot be factored out of the matrices $[H]$ and $[G]$. However, given the frequency parameter, i.e., the frequency of excitation ω and the speed of sound c, the standing pressure wave distribution and its normal derivative can be computed using equation (5.14). The analysis thus performed is designated as the "time-harmonic analysis".

On the contrary, in the eigenvalue analysis, we are required to compute the frequency parameter k, i.e., ω along with the characteristic pressure distribution in the acoustic enclosure. It is apparent that the boundary element formulation of equation (5.14) does not directly lend itself to an algebraic eigenvalue problem, as the frequency parameter k cannot be factored out of the matrices $[H]$ and $[G]$. In the subsequent chapters we will describe different approaches proposed in solving the eigenvalue problem.

The above formulation can be used in performing harmonic analysis for interior problems as well as problems with domains extending to infinity. The excitation may be caused by inhomogeneous boundary conditions in the case of radiation problems, or by wave patterns travelling from infinity and impinging on obstacles in the case of scattering problems. Often the excitation is caused by a vibrating structure, which interacts with the fluid medium adjacent to it. In this case, both the fluid and the structure must be considered together as a coupled fluid–structure system, which can be solved using a coupled BE–FE (finite element) hybrid formulation. The coupling of BE representing an infinite-extent fluid to an FE representation of a vibrating structure forms a powerful and frequently-used analysis approach. This procedure for coupling the fluid domain with the structure is presented below.

5.2.1. Acoustic fluid–structure interaction

Let us consider a vibrating structure adjacent to the acoustic fluid domain. Let us assume that the structure is discretized using the FEM thereby allowing structures with complex geometry and inhomogeneity to be modeled with relative ease. The matrix equation of a harmonically vibrating structure for solving structural degrees of freedom $\boldsymbol{d}$ is written as:

$$\left(-\omega^2[M] + \mathrm{j}\omega[C] + [K]\right)\{d\} = \{F_\mathrm{u}\} + \{F\} \tag{5.15}$$

$[M]$, $[C]$ and $[K]$ are the mass, damping and stiffness matrices of the discretized structure. $\{d\}$ is the nodal displacement vector consisting of the displacement components d_x, d_y and d_z for the 3-D problem. $\{F_\mathrm{u}\}$ is the pressure force exerted by the fluid on the structure at the fluid–structure interface and $\{F\}$ represents all other nodal forces that may be acting on the structure. The pressure force $\{F_\mathrm{u}\}$ at the nodes on the interface boundary of the structure is given by:

$$\{F_\mathrm{u}\} = \int_S \theta_i \, un \mathrm{d}S \tag{5.16}$$

θ_i are the shape functions employed in the finite element discretization of the structure. "n" is the normal, drawn outward from the fluid into the structure, at the interface boundary. Representing the pressure by shape functions and nodal pressure degrees of freedom ($u = \phi_j u_j$) and substituting for this pressure into the above equation yields the matrix equation for the pressure force at the interface:

$$\{F_u\} = [R]\{u\} \tag{5.17}$$

Matrix $[R]$ is formed by the usual finite element process where the integration is performed over each of the element surface that is on the interface boundary and summed using the finite element assembly algorithm:

$$[R] = \sum_{\mathrm{e}} \int_{S_\mathrm{e}} \theta_i \phi_j \, n \mathrm{d}S \tag{5.18}$$

Using equation (5.17), the finite element matrix equation (5.15) of the structure is rewritten as:

$$\left(-\omega^2[M] + \mathrm{j}\omega[C] + [K]\right)\{d\} - [R]\{u\} = \{F\} \tag{5.19}$$

The fluid pressure gradient is related to the structural displacement by the following equilibrium equation at the fluid–structure interface:

$$\left\{\frac{\partial u}{\partial n}\right\} = \rho\omega^2 \{d_\mathrm{n}\} \tag{5.20}$$

$\{d_\mathrm{n}\}$ is the normal displacement of the nodes at interface boundary and ρ is the fluid density. The normal displacement can be represented in its component directions as:

$$\{d_\mathrm{n}\} = \{n\}^T \begin{Bmatrix} d_x \\ d_y \\ d_z \end{Bmatrix} \tag{5.21}$$

$\{n\}^T = \langle n_x \; n_y \; n_z \rangle$, where n_x, n_y and n_z are the x, y and z components of the unit normal. In view of equations (5.20) and (5.21), equation (5.14) can be written as:

$$[H]\{u\} = \rho\omega^2[G]\{n\}^T \begin{Bmatrix} d_x \\ d_y \\ d_z \end{Bmatrix} \tag{5.22}$$

Using the notation $\{d\} = \langle d_x \; d_y \; d_z \rangle$, the acoustic boundary element matrix equation is written in a compact form:

$$[H]\{u\} = \rho\omega^2[G]\{n\}^T\{d\} \tag{5.23}$$

Now, combining equations (5.19) and (5.23) the coupled matrix equation is written as follows:

$$\begin{bmatrix} (-\omega^2[M] + \mathrm{j}\omega[C] + [K]) & -[R] \\ -\rho\omega^2 G n^T & [H] \end{bmatrix} \begin{Bmatrix} \{d\} \\ \{u\} \end{Bmatrix} = \begin{Bmatrix} \{F\} \\ \{0\} \end{Bmatrix} \tag{5.24}$$

The coupling of fluid and structure occurs through the off-diagonal sub-matrices $-[R]$ and $-\rho\omega^2[G]\{n\}^T$ in equation (5.24). This augmented matrix equation can either be solved simultaneously for the unknown vectors $\{d\}$ and $\{u\}$ [166–168], or one of the vectors can be eliminated to recast it as a matrix equation of a single unknown vector [169–172].

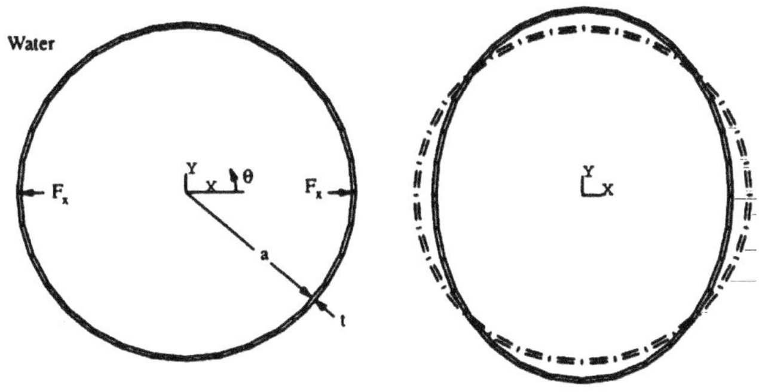

Figure 5.1. Cylindrical shell and mode investigated (shell: $a = 0.254$ m, $t = 0.00635$ m, Young's modulus $E = 2.068$ E11 N/m^2, density $\rho_s = 7929$ kg/m^3; water: speed of sound $C = 1460$ m/s, density $\rho_w = 1030$ kg/m^3). (a) Shell cross section; (b) Mode 2.

5.2.2. Example

Dynamic response of submerged cylindrical shell

A practical application of the acoustic fluid–structure interaction formulation described above is the study of the dynamics of structures submerged in fluid of infinite extent. Here we investigate the dynamic response of a circular cylindrical shell submerged in an infinite body of water [167]. Schroeder and Marcus [173] have presented the eigenfrequencies of the problem in plane 2-D assuming the shell to be infinitely long. The same dimensions and material properties of the shell used by them have been employed in the vibration response computations in order to validate the results obtained. The cross-section of the steel cylindrical shell is shown in Figure 5.1a. It is discretized using 32 finite elements around the circumference. Plane 2-D elements of the ANSYS program [160] have been used to model the shell. The elements representing plane strain conditions allow bilinear variation of the d_x and d_y displacements. The BE discretization of the external fluid domain consists of 32 nodes representing the fluid pressure on the outer surface of the shell. The shell displacement and fluid pressure are coupled in the manner described in the last section.

In order to excite the shell into its first fundamental circumferential mode, designated as mode 2 in reference 173, two diametrically opposing forces were applied as shown in Figure 5.1a. For frequencies in the neighborhood of mode 2 resonance, the radial displacement at the outer radius at $\theta = 0$ location is plotted in Figure 5.2 and the shell deformation configuration for mode 2 is shown in Figure 5.1b. For comparison, entirely FE solutions considering finite extents of water surrounding the ring with vacuum beyond are also plotted in Figure 5.2. In addition, the FE solution of the ring in vacuum is also plotted. Concentric fluid boundaries placed at radii $R = 0.559$ and 0.828 m are the two cases of finite fluid extent that were modeled by the linear acoustic fluid finite element of the ANSYS program. The same level of circumferential discretization as in the BE–FE solution was used. In the fluid domain, four and six radial divisions were used, respectively, for $R = 0.559$ and 0.828 m. Zero pressure was specified on the outer fluid boundary to represent vacuum beyond radius R. Alternatively, the outer boundary could have been modeled as zero pressure gradient boundary with "dashpots" connected to the fluid grid at this boundary to approximate the non-reflecting condition. In the present study, however, we use the zero pressure condition at

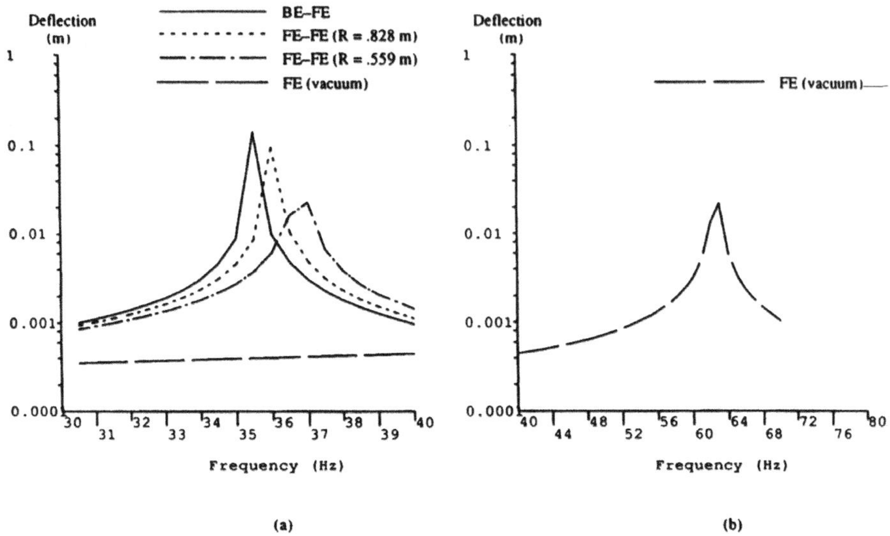

Figure 5.2. Response of the cylindrical shell (radial deflection at outer radius at $\theta = 0$). (a) Shell in water; (b) shell in vacuum.

Table 5.1. Mode 2 resonant frequencies of the cylindrical shell.

Shell configuration	Analytical resonant eigenfrequency (Hz)	Eigenfrequency from vibration response (Hz)	Computational method
In vacuum	62.39	62.50	FE
Finite fluid ($R = 0.559$ m)	36.80	36.75	FE–FE
Finite fluid ($R = 0.828$ m)	35.99	36.25	FE–FE
Infinite fluid	35.76	35.75	BE–FE

the outer boundary in order to compare results with the eigenfrequency solution found in reference 173. From the graphs in Figure 5.2, the resonant peaks of the submerged ring shift to the left as the radial extent of the fluid domain is increased in the acoustic FE solution. The response plotted from the BE–FE coupled formulation represents the solution to the problem as posed when the ring is submerged in a fluid extending to infinity. The resonant frequency obtained from this frequency response computation is $f = 35.75$ Hz, and it closely matches with the analytically found eigenfrequency of 35.76 Hz given by Schroeder and Marcus [173]. Table 5.1 shows the resonant frequencies for different configurations of the ring obtained from the frequency response sweep. The analytical eigenfrequency results are also given for comparison. Although the resonant frequency of the FE–FE solution has converged to within one per cent of the infinite fluid domain frequency, the response plots in Figure 5.2 show a much larger error in the displacement of the ring, especially close to the resonant frequency.

5.3. Elasticity

Here we will only discuss the solution of the time-harmonic or steady-state elastodynamic problem because it is most closely related to the free vibration analysis of structures. Consider a linear, elastic, homogeneous and isotropic domain Ω bounded by a surface Γ. The governing equation of motion is:

$$\nabla \cdot \boldsymbol{\sigma}(\boldsymbol{x}, t) + \rho \boldsymbol{b}(\boldsymbol{x}, t) = \rho \frac{\partial^2 \boldsymbol{u}(\boldsymbol{x}, t)}{\partial t^2} \tag{5.25}$$

where $\boldsymbol{\sigma}(\boldsymbol{x}, t)$ is the stress tensor, ρ is the mass density of the deformed body, $\boldsymbol{b}(\boldsymbol{x}, t)$ is the body force vector, $\boldsymbol{u}(\boldsymbol{x}, t)$ is the displacement vector and t is the time. Assuming infinitesimal deformations the kinematic relations are given by:

$$\boldsymbol{\varepsilon}(\boldsymbol{x}, t) = \frac{1}{2}\left[\nabla \boldsymbol{u} + \boldsymbol{u}\nabla\right] \tag{5.26}$$

$\boldsymbol{\varepsilon}(\boldsymbol{x}, t)$ is the strain tensor. The constitutive relation is the Hooke's law which can be written in terms of the Lamé's constants λ and μ:

$$\boldsymbol{\sigma} = \lambda \,\mathrm{tr}\, \boldsymbol{\varepsilon}\, \mathbf{1} + 2\mu \boldsymbol{\varepsilon} \tag{5.27}$$

$\mathrm{tr}\, \boldsymbol{\varepsilon}$ is the trace of $\boldsymbol{\varepsilon}$ and $\mathbf{1}$ is the unit tensor. The Lamé's constants λ and μ are related to the Poisson's ratio and Young's modulus in this manner:

$$\lambda = \frac{\nu E}{(1+\nu)(1-2\nu)} \tag{5.28}$$

$$\mu = G = \frac{E}{2(1+\nu)} \tag{5.29}$$

G is the shear modulus. Combining equations (5.26) and (5.27) the stress tensor $\boldsymbol{\sigma}$ can be expressed in terms of displacements $\boldsymbol{u}$:

$$\boldsymbol{\sigma} = \lambda \nabla \cdot \boldsymbol{u} \mathbf{1} + \mu(\nabla \boldsymbol{u} + \boldsymbol{u}\nabla) \tag{5.30}$$

Substituting this expression for the stress tensor into the equation of motion (5.25), we arrive at the following Navier–Cauchy equation of motion:

$$\mu \nabla^2 \boldsymbol{u} + (\lambda + \mu)\nabla\nabla \cdot \boldsymbol{u} + \rho \boldsymbol{b} = \rho \frac{\partial^2 \boldsymbol{u}}{\partial t^2} \tag{5.31}$$

Since we are only concerned with the time-harmonic elastodynamic problem, we assume:

$$\left.\begin{aligned} \boldsymbol{u}(\boldsymbol{x}, t) &= \boldsymbol{u}(\boldsymbol{x}, \omega) e^{-\mathrm{j}\omega t} \\ \boldsymbol{t}(\boldsymbol{x}, t) &= \boldsymbol{t}(\boldsymbol{x}, \omega) e^{-\mathrm{j}\omega t} \\ \boldsymbol{b}(\boldsymbol{x}, t) &= \boldsymbol{b}(\boldsymbol{x}, \omega) e^{-\mathrm{j}\omega t} \end{aligned}\right\} \tag{5.32}$$

where $\boldsymbol{u}(\boldsymbol{x}, \omega)$, $\boldsymbol{t}(\boldsymbol{x}, \omega)$ and $\boldsymbol{b}(\boldsymbol{x}, \omega)$ are the amplitudes of the displacement, traction and body force vectors respectively. By substituting the expressions of $\boldsymbol{u}(\boldsymbol{x}, t)$ and $\boldsymbol{b}(\boldsymbol{x}, t)$ from above equation into the equation of motion (5.31), we obtain the following the governing differential equation of the steady-state elastodynamics:

$$\mu \nabla^2 \boldsymbol{u}(\boldsymbol{x}, \omega) + (\lambda + \mu)\nabla\nabla \cdot \boldsymbol{u}(\boldsymbol{x}, \omega) + \rho\omega^2 \boldsymbol{u}(\boldsymbol{x}, \omega) = -\rho \boldsymbol{b}(\boldsymbol{x}, \omega) \tag{5.33}$$

which is the vector Helmholtz equation. The solution of this equation must satisfy specified boundary conditions on the boundary Γ of the domain Ω:

$$\left.\begin{aligned} \boldsymbol{u}(\boldsymbol{x}, \omega) &= \boldsymbol{u}_b(\boldsymbol{x}, \omega) \\ \boldsymbol{t}(\boldsymbol{x}, \omega) &= \boldsymbol{\sigma}(\boldsymbol{x}, \omega)\boldsymbol{n}(\boldsymbol{x}) = \boldsymbol{t}_b(\boldsymbol{x}, \omega) \end{aligned}\right\} \tag{5.34}$$

where $\boldsymbol{n}(\boldsymbol{x})$ is the unit normal vector at the boundary Γ. The fundamental solution for the governing differential equation (5.33), $\boldsymbol{u}^*(p, q, \omega)$, can be found by solving this equation due a unit harmonic body force. Thus, $\boldsymbol{u}^*(p, q, \omega)$ is the solution of:

$$\mu \nabla^2 \boldsymbol{u}^*(p, q, \omega) + (\lambda + \mu)\nabla\nabla \cdot \boldsymbol{u}^*(p, q, \omega) + \rho\omega^2 \boldsymbol{u}^*(p, q, \omega) = -\boldsymbol{\delta}(p, q) \tag{5.35}$$

where $\boldsymbol{\delta}(p, q)$ is the Kronecker delta symbol and (p, q) are the source and field points inside the domain. The fundamental solutions in this case are found to be:

$$u^*_{ij}(p, q, \omega) = \frac{\mathrm{j}}{4\rho}\left[\frac{1}{c_s^2}\delta_{ij}H_0(k_s r) - \frac{1}{\omega^2}\frac{\partial^2}{\partial x_i \partial x_j}\{H_0(k_p r) - H_0(k_s r)\}\right] \quad \text{(2-}D\text{)} \tag{5.36}$$

$$u^*_{ij}(p, q, \omega) = \frac{1}{4\pi\rho\omega^2}\left[\delta_{ij}k_s\frac{e^{\mathrm{j}k_s r}}{r} - \frac{\partial^2}{\partial x_i \partial x_j}\left(\frac{e^{\mathrm{j}k_p r}}{r} - \frac{e^{\mathrm{j}k_s r}}{r}\right)\right] \quad \text{(3-}D\text{)} \tag{5.37}$$

where $H_0(\ )$ is the Hankel function of zero-order and first kind, $(k_p = \omega/c_p,\ k_s = \omega/c_s)$ are the wave numbers for the P-wave (or dilatational or irrotational wave) and S-wave (or shear or isochoric wave) and (c_p and c_s) are the P-wave and S-wave velocities respectively. c_p and c_s are given by:

$$c_p = \sqrt{\frac{\lambda + 2\mu}{\rho}} \tag{5.38}$$

$$c_s = \sqrt{\frac{\mu}{\rho}} \tag{5.39}$$

Using the fundamental solutions (5.36) and (5.37), the boundary element equation, relating the displacements and tractions at the boundary Γ, can be developed:

$$\begin{aligned} &C_{ij}u_i(P, \omega) + \int_\Gamma T^*_{ij}(P, Q, \omega)u_i(Q, \omega)\,\mathrm{d}\Gamma(Q) \\ &= \int_\Gamma t_i(Q, \omega)u^*_{ij}(P, Q, \omega)\,\mathrm{d}\Gamma(Q) + \int_\Omega \rho b_i(p, \omega)u^*_{ij}(p, q, \omega)\,\mathrm{d}\Omega \end{aligned} \tag{5.40}$$

where T^*_{ij} is the surface traction at Q due to unit body force. The value of T^*_{ij} can be found as follows: substitute the expression for the displacements from equations (5.36) or (5.37) into equation (5.30) to compute stress due to unit body force. Substitute this stress into the second part of equation (5.34) to compute T^*_{ij}. As in the case of acoustic harmonic analysis, given the wave velocities and frequency of vibration, the harmonic vibration of the structure can be found from the boundary conditions [eqns. (5.34)] using the above equation. Once again, if we were to perform free vibration analysis of the structure, where we are required to compute the frequencies of vibration, it could not be done in a straightforward way. This is because the frequency parameters, k_p and k_s, are deeply embedded into transcendental functions of this equation and so the equation cannot be cast as an algebraic eigenvalue problem.

Chapter 6

Dynamic Analysis: Acoustics and Elasticity

6.1. Introduction

So far we have discussed steady-state and time-harmonic analysis in acoustics and elasticity. This chapter will deal with the time-dependent equations as applied to boundary element method (BEM). Unlike in finite element method (FEM), the transient problem in BEM requires special treatment of the inertial effects introduced by the transient terms of the governing differential equation. This section will introduce the concept of mass matrix in the context of BEM and as such set the stage for the formulation of algebraic eigenvalue problems in the next chapter. It will become evident that the idea of stiffness and mass matrices does not show up in a straightforward fashion in the BEM. The dynamic equations in acoustics as well as in elasticity will be dealt with in this section.

The dynamic governing equation [eqn. (5.1)] in acoustics is rewritten again as a starting point to show the boundary element discretization in dynamics analysis:

$$\nabla^2 \Psi = \frac{1}{c^2} \frac{\partial^2 \Psi}{\partial t^2} \tag{6.1}$$

There are two ways of dealing with the solution to the above transient equation: (a) time-dependent Green's function method and (b) static Green's function method. Even though in the literature several other specific names are utilized, we will use the above designations to help develop the concepts behind these methods.

Time-dependent Green's function method

Here we use a time-dependent Green's function to the equation as the weighting function and develop the boundary discretization as outlined in Chapter 2. This method has been used quite extensively to solve for the unknowns as a function of time [see for example reference 71 and works cited there]. However, in this method there are no distinctly identifiable mass or stiffness matrices and as such is not extensible to time-harmonic case. For example, in the FEM, the discretized dynamic equation for a conservative system is given as follows:

$$[K]\{\Psi\} + [M]\{\ddot{\Psi}\} = \{0\} \tag{6.2}$$

The time integration on this equation can be performed to obtain the time-dependent solution $\{\Psi\}$. If the excitation is time-harmonic, the same equation can be easily transformed into a quasi-steady state equation using the substitution $\{\Psi\} = \{\overline{\Psi}\}e^{j\omega t}$. The same equation will also be applicable to an eigenvalue analysis. The time-dependent

Green's function method in the boundary element, however, does not allow this type of flexibility.

Static Green's function method

In this method, we make use of a static Green's function, i.e., Green's function corresponding to the first term(s) (without the inertia term) of the governing equation, as the weighting function and develop the boundary discretization of the first term(s). The volume integral for the second or inertia term is retained and requires further treatment to transform it into boundary-only integrals. This inertial term is eventually transformed into a mass matrix leading to a discretized equation similar to equation (6.2). This method is attractive because of its simplicity in application to pure transient, time-harmonic or eigenvalue analysis. Since the subsequent chapters of this book are primarily devoted to describing eigenvalue analysis in boundary element, the static Green's function method is briefly introduced here. Section 6.1.1 will use the wave equation (6.1) to describe this method, whereas Section 6.1.2 will show the application of the formulation to elastodynamic equation.

6.1.1. Static Green's function method in acoustics

The first term of the governing dynamic equation (6.1) can be converted into a boundary integral statement using the static Green's function, given in equation (2.8) or (2.9):

$$C_p p_p + \int_\Gamma p q^* \,\mathrm{d}\Gamma - \int_\Gamma q p^* \,\mathrm{d}\Gamma + \frac{1}{c^2}\int_\Omega p^* \ddot{p}\,\mathrm{d}\Omega = 0 \tag{6.3}$$

The inertia term is left intact. This term involves integration over the volume of the entire domain. In order to achieve a boundary-only formulation, the volume integral needs to be transformed into boundary integrals. There are two methods that achieve this transformation. They are: Dual Reciprocity Method (DRM) and Particular Integral Method (PIM). Both these methods lead to the same result. Here we will use DRM to outline the transformation. According to this method, a function Θ must be found such that

$$\nabla^2\Theta = \ddot{p}\,(x,t) \tag{6.4}$$

inside the domain. Then the inertial volume integral term in equation (6.3) can be transformed into boundary-only integrals with the help of Gauss's divergence theorem. Consequently, equation (6.3) will become:

$$[G]\{q] - [H]\{p\} = (1/c^2)(-[G]\{\partial\Theta/\partial n\} + [H]\{\Theta\}) \tag{6.5}$$

Since the divergence theorem is applied for the second time (this time only to the inertial term) to arrive at equation (6.5), the method is given the name "DRM" [120]. The unknown function Θ, contained in the inertial term of equation (6.5), is related to the unknown pressure p in the domain through the differential equation (6.4). We can approximate the unknown variable $\ddot{p}$ in the inertial term by a global shape function in the following manner:

$$\ddot{p}\,(x,t) = \sum_{m=1}^{\infty} f(x,\xi_m)\,\phi_m(t) \tag{6.6}$$

where x is a point in the domain, ξ_m is a source boundary point m and ϕ is a fictitious density function at Γ_m. The functions $f(x,\xi_m)$ are called global shape functions which are used to interpolate p. The specific forms of these global shape functions that have been employed successfully will be discussed in the subsequent chapters when we talk about the algebraic eigenvalue formulations. The next step is to substitute equation (6.6) into (6.4):

$$\nabla^2\Theta = \sum_{m=1}^{\infty} f(x, \xi_m)\, \phi_m(t) \tag{6.7}$$

This can be integrated to solve for Θ. The integration can be performed by the method of undetermined coefficients or by trial and error once the global shape function $f(x, \xi_m)$ is chosen. The resulting solution for Θ can then be written as:

$$\Theta == \sum_{m=1}^{\infty} g(x, \xi_m)\, \phi_m(t) \tag{6.8}$$

The normal derivative of Θ can then be derived as:

$$\partial\Theta/\partial n == \sum_{m=1}^{\infty} \frac{\partial g}{\partial n}(x, \xi_m)\, \phi_m(t) \tag{6.9}$$

Taking the discretized boundary points to be the interpolation points, equations (6.6), (6.8) and (6.9) can be written in matrix form:

$$\{\ddot{p}\,(x,t)\} = [F]\{\phi(t)\} \tag{6.10}$$

$$\{\Theta\} = [D]\{\phi(t)\} \tag{6.11}$$

$$\{\partial\Theta/\partial n\} = [E]\{\phi(t)\} \tag{6.12}$$

We can now substitute the matrix equations (6.11) and (6.12) into equation (6.5) to arrive at:

$$[G]\{q\} - [H]\{p\} = (1/c^2)(-[G][E] + [H][D])\{\phi(t)\} \tag{6.13}$$

Solving for $\{\phi(t)\}$ from equation (6.10) and substituting back into (6.13), we arrive at the following discretized dynamic equation:

$$[G]\{q\} - [H]\{p\} = (1/c^2)(-[G][E] + [H][D])[F]^{-1}\{\ddot{p}\} \tag{6.14}$$

which can be written in the following compact form:

$$[G]\{q\} - [H]\{p\} = (1/c^2)[\overline{M}]\{\ddot{p}\} \tag{6.15}$$

where

$$[\overline{M}] = (-[G][E] + [H][D])[F]^{-1} \tag{6.16}$$

is the desired mass matrix for the dynamic problem. For convenience of illustration we multiply both sides of equation (6.15) by $[G]^{-1}$:

$$[K]\{p\} + (1/c^2)[M]\{\ddot{p}\} = \{q\} \tag{6.17}$$

where $[K] = [G]^{-1}[H]$ and $[M] = [G]^{-1}[\overline{M}]$. In a given problem where pressure or pressure gradient boundary conditions are known on the boundary, the equation (6.17) can be partitioned. If we assume that $\{p_1\}$ is specified on Γ_1 and $\{q_2\}$ is specified on Γ_2, the partitioned form of equation (6.17) becomes:

$$\begin{bmatrix} K_{11} & K_{12} \\ K_{21} & K_{22} \end{bmatrix} \begin{Bmatrix} p_1 \\ p_2 \end{Bmatrix} + \frac{1}{c^2} \begin{bmatrix} M_{11} & M_{12} \\ M_{21} & M_{22} \end{bmatrix} \begin{Bmatrix} \ddot{p}_1 \\ \ddot{p}_2 \end{Bmatrix} = \begin{Bmatrix} q_1 \\ q_2 \end{Bmatrix} \tag{6.18}$$

In the above equation, $\{p_2\}$ and $\{q_1\}$ are unknown quantities to be solved for. Note that there are no time derivatives of pressure gradient $\{q\}$ in these dynamic equations. As a result, we can recast the dynamic equation in terms of $\{p_2\}$ alone using only the second equation from (6.18):

$$[K_{22}]\{p_2\} + (1/c^2)[M_{22}]\{\ddot{p}_2\} = \{q_2\} - [K_{21}]\{p_1\} - (1/c^2)[M_{21}]\{\ddot{p}_1\} \tag{6.19}$$

This is now a standard transient dynamic matrix equation, which can be solved for the unknown pressure $\{p_2\}$ as a function of time given the initial conditions on $\{q_2\}$ and $\{p_1\}$.

6.1.2. Static Green's function method in elasticity

Here we start from the governing equation [eqn. (5.31)], which is the elastodynamic equilibrium equation in terms of displacements in the absence of body forces:

$$\mu u_{i,jj} + (\lambda + \mu) u_{j,ji} = \rho \ddot{u}_i \tag{6.20}$$

μ is the shear modulus and λ is the Lamé's constant:

$$\mu = E/[2(1+\nu)] \tag{6.21a}$$

$$\lambda = (\nu E)/[(1+\nu)(1-2\nu)] \tag{6.21b}$$

ν is the Poisson's ratio and ρ is the density. The boundary integral equation for the left-hand side of equation (6.20) using the static fundamental solution u^* is given by:

$$C_{ij} u_j - \int_\Gamma t^*_{ij} u_j \, d\Gamma + \int_\Gamma t_j u^*_{ij} \, d\Gamma = \rho \int_\Omega u^*_{ij} \ddot{u}_j \, d\Omega \tag{6.22}$$

where u^* and t^* are the Kelvin's displacement and traction fundamental solutions given as:

For (2-D):

$$u^*_{ij} = \frac{1}{8\pi\mu(1-\nu)} \left[(3-4\nu)\delta_{ij} \ln\left(\frac{1}{r}\right) r_{,i} r_{,j} \right] \tag{6.23}$$

$$t^*_{ij} = \frac{-1}{4\pi r(1-\nu)} \left[\{(1-2\nu)\delta_{ij} + 2r_{,i} r_{,j}\} \frac{\partial r}{\partial n} - (1-2\nu)\left(r_{,i} n_j - r_{,j} n_i\right) \right] \tag{6.24}$$

For (3-D):

$$u^*_{ij} = \frac{1}{16\pi\mu r(1-\nu)} \left[(3-4\nu)\delta_{ij} + r_{,i} r_{,j} \right] \tag{6.25}$$

$$t^*_{ij} = \frac{-1}{8\pi r^2(1-\nu)} \left[\{(1-2\nu)\delta_{ij} + 3r_{,i} r_{,j}\} \frac{\partial r}{\partial n} - (1-2\nu)\left(r_{,i} n_j - r_{,j} n_i\right) \right] \tag{6.26}$$

As shown in the previous section, the volume integral in equation (6.22) can be transformed into a boundary-only integral employing the DRM. To this end, we first propose to find a function U_i (analogous to Θ in the previous section, eqn. (6.4)) which would satisfy right-hand side of the governing dynamic equation (6.20):

$$\mu U_{i,jj} + (\lambda + \mu) U_{j,ji} = \ddot{u}_i \tag{6.27}$$

We substitute this value of $\ddot{u}_i$ into the right-hand side of equation (6.22) and apply Gauss's divergence theorem to the right-hand side to arrive at:

$$C_{ij} u_j - \int_\Gamma t^*_{ij} u_j \, d\Gamma + \int_\Gamma t_j u^*_{ij} \, d\Gamma = -\rho \left[\int_\Gamma t^*_{ij} U_j d\Gamma + \int_\Gamma T_i u^*_{ij} \, d\Gamma \right] \tag{6.28}$$

where T_i are the traction components corresponding to U_i. The acceleration $\ddot{u}_i$ in equation (6.27) is approximated by a product of global shape functions $f_i(x, \xi_m)$ and a fictitious density function ϕ_i:

$$\ddot{u}_i = \sum_{m=1}^{\infty} f_i(x, \xi_m) \, \phi_m(t) \tag{6.29}$$

Various forms of global shape functions $f_i(x, \xi_m)$ can be chosen. We will come back to the discussion of suitability of different shape function in subsequent chapters. Once these global shape functions are selected, we substitute the expression for $\ddot{u}_i$ from equation (6.29) into equation (6.27). The functions U_i can then be determined from equation (6.27) by the method of undetermined coefficients or by trial and error:

$$U_i = \sum_{m=1}^{\infty} g_i(x, \xi_m) \, \phi_m(t) \tag{6.30}$$

The traction components T_i corresponding to U_i can be found using the compatibility equations, the constitutive relations and the traction–stress relations. These relations are written here for completeness.

- The compatibility equations:

$$\varepsilon_{ij} = \frac{1}{2} \left\{ \frac{\partial U_i}{\partial x_j} + \frac{\partial U_j}{\partial x_i} \right\} \tag{6.31}$$

- The constitutive relations:

$$\sigma_{ij} = \frac{2\mu v}{(1 - 2v)} \delta_{ij} \varepsilon_{kk} + 2\mu \varepsilon_{ij} \tag{6.32}$$

- The traction–stress relations:

$$T_i = \sigma_{ij} n_j \tag{6.33}$$

Using equation (6.30) in the above set of equations, the traction components T_i can be written as:

$$T_i = \sum_{m=1}^{\infty} \hat{g}_i(x, \xi_m) \, \phi_m(t) \tag{6.34}$$

After discretization of the boundary, the boundary integral equation (6.28) can be transformed into a set of matrix equations:

$$[H]\{u\} - [G]\{t\} = \rho([G]\{T\} - [H]\{U\}) \tag{6.35}$$

Taking the discretized boundary points to be the interpolation points for the approximation in equation (6.29), equations (6.29), (6.30) and (6.34) can be written in the matrix form as follows:

$$\{\ddot{u}\} = [F]\{\phi\} \tag{6.36}$$

$$\{U\} = [D]\{\phi\} \tag{6.37}$$

$$\{T\} = [E]\{\phi\} \tag{6.38}$$

Utilizing these three equations, the discretized dynamic boundary element equation (6.35) can be written as:

$$[H]\{u\} - [G]\{t\} = \rho([G][E] - [H][D])[F]^{-1}\{\ddot{u}\} \tag{6.39}$$

Let us write this equation as:

$$[H]\{u\} - [G]\{t\} = \rho[\overline{M}]\{\ddot{u}\} \tag{6.40}$$

where:

$$[\overline{M}] = ([G][E] - [H][D])[F]^{-1} \tag{6.41}$$

As before, we multiply both sides of equation (6.40) by $[G]^{-1}$ to obtain:

$$[K]\{u\} - [M]\{\ddot{u}\} = \{t\} \tag{6.42}$$

where $[K]$ is the stiffness matrix:

$$[K] = [G]^{-1}[H] \tag{6.43}$$

and $[M]$ is the mass matrix:

$$[M] = \rho[G]^{-1}[\overline{M}] \tag{6.44}$$

Written in this form the discretized dynamic equation (6.42) is similar to the finite element equation. We now assume that $\{u\} = \{\bar{u}\}$ on the portion of the boundary Γ_1 and $\{t\} = \{\bar{t}\}$ on the other portion of the boundary Γ_2 are the given boundary conditions. Using these boundary divisions, we can partition the equation (6.42) into the following:

$$\begin{bmatrix} K_{11} & K_{12} \\ K_{21} & K_{22} \end{bmatrix} \begin{Bmatrix} u_1 \\ u_2 \end{Bmatrix} + \begin{bmatrix} M_{11} & M_{12} \\ M_{21} & M_{22} \end{bmatrix} \begin{Bmatrix} \ddot{u}_1 \\ \ddot{u}_2 \end{Bmatrix} = \begin{Bmatrix} t_1 \\ t_2 \end{Bmatrix} \tag{6.45}$$

Once again, we note that the dynamic equations do not involve terms containing the time derivatives of the traction and therefore the problem can be posed in terms of the unknowns $\{u_2\}$ from the bottom row of equations (6.45) as follows:

$$[K_{22}]\{u_2\} + [M_{22}]\{\ddot{u}_2\} = \{t_2\} - [K_{21}]\{u_1\} - [M_{21}]\{\ddot{u}_1\} \tag{6.46}$$

This is now a standard transient dynamic matrix equation for structural dynamics, which can be solved for the unknown displacement $\{u_2\}$ as a function of time given the initial conditions on $\{t_2\}$ and $\{u_1\}$.

6.2. Eigenvalue problem in acoustics

Often when dealing with acoustic cavities, the resonant frequencies of the enclosed domain are of practical interest. For example, knowledge of the acoustic resonant frequencies is essential in the design of an automobile passenger cabin in order to reduce noise levels. Also, in the design of an auditorium to improve the sound quality perceived by the audience, the acoustic design process requires knowledge of the resonant frequencies of the auditorium.

The computation of the resonant frequencies involves the free response calculation, starting from the acoustic wave equation (5.1). For the harmonic oscillations of the acoustic pressure, the homogeneous Helmholtz equation (5.2) is written here once again

$$\nabla^2 u + k^2 u = 0 \tag{6.47}$$

We are interested in the response frequencies ($k = \omega/c$) and mode shapes of the pressure (u) within an acoustic cavity, such as an auditorium. In other words, the intent is to identify the characteristic mode shapes and the frequencies that are unique to a particular cavity. In finite element (FE) discretized formulations, the following algebraic eigenvalue problem arises naturally from equation (6.47)

$$[K]\{u\} = k^2[M]\{u\} \tag{6.48}$$

where $[K]$ and $[M]$ are "stiffness" and "mass" matrices resulting from the Laplacian term $\nabla^2 u$ and the inertia term $k^2 u$, respectively. In boundary element (BE) discretizations, additional steps are required to arrive at an algebraic equation equivalent to equation (6.48).

Proceeding from equation (6.47), in the same manner as in Section 5.2, we arrive at the boundary discretized Helmholtz equation (5.14), which is given here once again

$$[H(k)]\{u\} = [G(k)]\{v\} \tag{6.49}$$

where the matrices $[H]$ and $[G]$ are shown explicitly as functions of the wave number k. As developed in Section 5.2, this is the same equation used to compute harmonic response to a time-harmonic force, such as, an oscillating pressure or flow. Therefore, the time-harmonic analysis involves the application of non-homogeneous boundary conditions, either a known excitation pressure u, or a flow v. This leads to a linear set of equations having a right-hand side load vector. Given the frequency of oscillation, ω, this set of equations can be solved to get the harmonic response pressure. However, in free response calculation, the right-hand load vector is $\{0\}$ resulting from homogeneous boundary conditions, namely, either $u = 0$, or $v = 0$:

$$[A(k)]\{x\} = \{0\} \tag{6.50}$$

Most acoustic eigenproblems involve acoustic fluid surrounded by rigid boundaries, in which case $v = 0$ boundary condition will apply for the entire boundary. As a result, in equation (6.50) the matrix $[A(k)] = [H(k)]$. In order for the vector $\{x\} \neq \{0\}$, the $\det[A(k)]$ must be 0. The characteristic values of the polynomial equation

$$\det[A(k)] = 0 \tag{6.51}$$

are the resonant frequencies k_i sought for. Posed in this fashion, the only way to find k_i is to employ a search method starting from an arbitrarily picked value for k_i and evaluating the determinant since each coefficient of matrix $[A]$ is a function of the frequency parameter k_i. This process of computing the resonant frequencies is termed as the Determinant Search Method (DSM), which will be described in Section 6.4.

6.3. Eigenvalue problem in elasticity

The dynamics of elastic structures can be characterized by the natural frequencies of vibration. The free vibration response calculation is routinely performed in the design of elastic structures. In order to compute the natural frequencies from the BEM equation, we start from the time-harmonic BE equations developed in Section 5.3. For convenience this boundary element equation [eqn. (5.40)] is rewritten here:

$$
\begin{aligned}
C_{ij}u_i(P,\omega) &+ \int_\Gamma T^*_{ij}(P,Q,\omega)u_i(Q,\omega)\,\mathrm{d}\Gamma(Q) \\
&= \int_\Gamma t_i(Q,\omega)u^*_{ij}(P,Q,\omega)\,\mathrm{d}\Gamma(Q) + \int_\Omega \rho b_i(p,\omega)u^*_{ij}(p,q,\omega)\,\mathrm{d}\Omega \qquad (6.52)
\end{aligned}
$$

After boundary discretization, this equation can be cast into matrix form:

$$[H(k_\mathrm{p},k_\mathrm{s})]\{u\} = [G(k_\mathrm{p},k_\mathrm{s})]\{t\} \qquad (6.53)$$

where $\{u\}$ and $\{t\}$ are the nodal displacement and traction vectors and $k_\mathrm{p}=\omega/C_\mathrm{p}$ and $k_\mathrm{s}=\omega/C_\mathrm{s}$, C_p and C_s being the dilatational and shear wave velocities in the elastic medium. After applying the boundary conditions [eqn. (5.34)], the above equation can be transformed into thc samc form as cquation (6.50). For a complctcly frcc structurc without any essential boundary conditions, the entire traction vector $\{t\}$ will be zero leading to:

$$[H(k_\mathrm{p},k_\mathrm{s})]\{u\} = \{0\} \qquad (6.54)$$

Unlike in acoustics, elastic structural problems with mixed boundary conditions are also of importance where on part of the boundary traction $t=0$ and the essential boundary conditions ($u=0$) is specified on the rest of the boundary. This leads to an equation similar to equation (6.51):

$$\det[A(k_\mathrm{p},k_\mathrm{s})]=0 \qquad (6.55)$$

Once again, in order to compute the natural frequencies of vibration, one has to resort to a search method, since the coefficients of matrix $[A]$ are functions of the parameters k_p and k_s which are in turn functions of ω.

6.4. Characteristic equation for eigenvalues

The equation (6.51) or (6.55) is known as the characteristic equation. It has multiple roots, which are the resonant frequencies of the system under consideration. The eigenvalue analysis procedure based on the FEM resorts to matrix algebra rather than dealing with the characteristic equation directly. Since the matrix $[A(\omega)]$ implicitly contains the frequency ω in it, matrix algebraic procedures are not generally applicable to the equations (6.51) or (6.55). Later in the book we will present formulations which will allow us to recast the BE eigenvalue problem in algebraic form. However, the next section will attempt to deal with the equations (6.51) or (6.55) directly.

6.4.1. Determinant search method

The matrix $[A]$, developed above, implicitly contains the resonant frequency ω in it and cannot be factored out to set up either a standard or a generalized algebraic eigenvalue problem. The standard and generalized eigenvalue problems require the matrices to be cast in the following forms:

$$[A]\{x\} = \omega\{x\} \qquad \text{(standard)} \tag{6.56}$$

$$[A]\{x\} = \omega[B]\{x\} \qquad \text{(generalized)} \tag{6.57}$$

where $[A]$ and $[B]$ are stiffness-type and mass-type matrices respectively and are independent of the frequency ω. However, the frequency ω is non-linearly embedded in the matrix $[A]$ of the equations (6.51) and (6.55), and as such a form of determinant search method (DSM) must be employed to determine the eigenvalues of the equation systems (6.51) and (6.55).

DSM is based on the complex-valued point load fundamental solutions given in equations (5.8) and (5.10) in acoustics and equations (5.36) and (5.37) in elastodynamics, giving rise to a complex determinant as equations (6.51) and (6.55). By arbitrarily choosing a value for ω, the complex valued determinant is evaluated in an iterative loop until both the real and imaginary parts go to zero.

$$\mathrm{Re}(\det[A]) = 0 \tag{6.58}$$

$$\mathrm{Im}(\det[A]) = 0 \tag{6.59}$$

The computational effort is enormous, since the search involves evaluation of complex determinants comprising of Gaussian elimination in complex arithmetic.

The real-valued determinant can be obtained by an approximation method in constructing the frequency domain equations that we started with equations (6.47) and (6.52). For example, if we use a real-valued particular solution of equation (6.47) as the Green's function, then we arrive at the determinant equation (6.51) where the coefficients in the matrix $[A(\omega)]$ will all be real-valued. Either the real or the imaginary component of the original complex Green's function for the governing differential equation can be used as the real-valued particular solution for this purpose. Thus, the determinant search will be limited to real arithmetic resulting in considerable savings of computational effort. The solution accuracy, however, will be approximate. So, this method may be used to get an estimate of the starting values of ω's to be used in the complex determinant search.

The DSM suffers from two major drawbacks:

(a) It requires the system matrix, which is complex-valued, to be formed repeatedly for different values of frequencies. This makes the technique extremely inefficient; and
(b) It is prone to failure in the case of closely spaced frequencies.

It should be apparent that DSM would be applied only when no other alternative is available. In the next section and Chapters 7 through 10 more efficient methods of BEM eigenvalue formulations are presented.

Example 6.1: Eigenfrequency calculation of a 2-D circular acoustic domain

The DSM presented in this section is applied to solve for the lowest resonant frequency of an acoustic circular domain of unit radius with soft boundary, i.e., $u = 0$ on Γ. This

Table 6.1. Lowest resonant frequency of a 2-D circular acoustic domain (closed-form solution $k_1 = 2.40482$) using complex fundamental solution.

Number of boundary segments	Resonant frequency computation from $\mathrm{Re}(\det A[k]) = 0$		Resonant frequency computation from $\mathrm{Im}(\det A[k]) = 0$	
	Resonant frequency	% Error in resonant frequency	Resonant frequency	% Error in resonant frequency
10	2.458	2.2	2.424	0.8
20	2.426	0.88	2.422	0.71
30	2.4176	0.53	2.4166	0.49
40	2.4142	0.39	2.4136	0.36
50	2.4121	0.3	2.4119	0.29

Table 6.2. Lowest resonant frequency of a 2-D circular acoustic domain (closed-form solution $k_1 = 2.40482$) using real-valued fundamental solution.

Number of boundary segments	Resonant frequency computation with $J_0(k)$ as the Green's function		Resonant frequency computation $Y_0(k)$ as the Green's function	
	Resonant frequency	% Error in resonant frequency	Resonant frequency	% Error in resonant frequency
20	2.4527	2	2.4247	0.82
30	2.4305	1.1	2.4147	0.51
40	2.4209	0.66	2.4138	0.37
80	2.4101	0.22	2.4093	0.18

type of problem has physical applications: (i) it can be used to determine the cut-off frequencies and eigenmode expansion in acoustic waveguides with soft walls; (ii) the Dirichlet eigenfunctions describe the TM modes in electromagnetic waveguides; (iii) the mechanical vibrations of membrane under tension lead to similar eigenvalue problem. DeMay [103, 104] solved this problem using BEM with constant elements. In reference [103] DeMay used the complex fundamental solution [eqn. (5.8)] as the Green's function. The closed-form solution for the lowest resonant frequency in this case is $k_1 = 2.40482$. The convergence study done by DeMay is shown in Table 6.1 for increasing number of boundary segments.

DeMay solved the same problem in reference [104] using real-valued fundamental solution. The results are presented in Table 6.2. Solutions using two real-valued fundamental solutions were generated. In one case the real part $J_0(k)$ and in the other case the imaginary part $Y_0(k)$ of the complex fundamental solution (Hankel function) to the Helmholtz equation was taken as the Green's function.

It appears from the results in Tables 6.1 and 6.2 that a real-valued fundamental solution may be adequate for resonant frequency computation. At the same time, the solution time will be considerably reduced. However, as mentioned before, the determinant must be computed repeatedly, as it contains the frequency parameter k in it. Moreover, DSM remains an inefficient error-prone method, especially for closely spaced resonant frequencies.

6.4.2. Enhanced determinant search method

The DSM can be enhanced to reduce computational effort in evaluating $\det[A(\omega)]$ for each assumed search value of ω. The idea is to expand the equations (6.50) or (6.54) in a Taylor series. Considering equation (6.50), the series expansion in ω becomes:

$$[A(\omega)] = [A_0] + \omega[A_1] + \omega^2[A_2] + \cdots + \omega^n[A_n] \quad \text{for real } \omega \tag{6.60}$$

The corresponding characteristic equation for the determination of eigenvalues can then be cast in the following form:

$$det([A_0] + \omega[A_1] + \omega^2[A_2] + \cdots + \omega^n[A_n]) = 0 \tag{6.61}$$

Note that the matrices $[A_0], [A_1], \ldots, [A_n]$ do not contain the frequency ω in them and, therefore, need not be formed at each determinant search loop. This solution technique may be looked upon as an enhanced determinant search method. It may be mentioned that this type of series expansion technique may be used in the time-harmonic analysis as well where the frequency sweep is used for response computation.

Chapter 7

Basics of Algebraic Eigenvalue Problem Formulation

7.1. Introduction

This chapter directly deals with the main idea of the book. Here we shall develop the algebraic eigenvalue formulations in the boundary element method (BEM), which have been a highly researched area in recent years. Even though the algebraic eigenvalue formulation in BEM may appear involved, it leads to an elegant and simple computational methodology. In the sections that follow we shall develop the earliest boundary element (BE) algebraic eigenvalue formulation which combines the BE formulation with finite element discretization. The finite element discretization of the domain is utilized specifically to formulate the mass matrix. The method, which is designated here as the Internal Cell Method (ICM), is developed first using the acoustics Helmholtz equation (6.1). Examples of the application of ICM are presented in which the natural frequencies and modes of vibration of plates are computed [113]. The subsequent chapters will deal with more recent BE algebraic eigenvalue formulations such as Dual Reciprocity and Particular Integral Methods (DRM and PIM).

7.2. Development of BE algebraic eigenvalue problem

The main idea of algebraic eigenvalue formulation in the BEM stems from the need to form a distinct mass matrix as in the finite element method (FEM). As we have seen in Chapter 6, the so-called static Green's function method would be the basis to form such a mass matrix. In the traditional formulation of BEM, the concept of mass matrix is not apparent. For example, the acoustic dynamic equation using time-dependent Green's function is given by:

$$C_p(\xi)p_p(\xi, t_F) + \int_{t_0}^{t_F} \int_\Gamma p(x,t)q^*(\xi, x, t_F, t)\,\mathrm{d}\Gamma(x)\,\mathrm{d}t$$

$$= \int_{t_0}^{t_F} \int_\Gamma q(x,t)p^*(\xi, x, t_F, t)\,\mathrm{d}\Gamma(x)\,\mathrm{d}t - \frac{1}{c^2}\int_\Omega \left\{ p_0(x, t_0)\left[\frac{\partial p^*(\xi, x, t_F, t)}{\partial t}\right]_0\right.$$

$$\left. - \left[\frac{\partial p(x,t)}{\partial t}\right]_0 p^*(\xi, x, t_F, t)\right\} d\Omega(x) \tag{7.1}$$

The matrix form of this equation is:

$$[H]\{p(t_F)\} = [G]\{q(t_F)\} + [B]\{p(t_{F-1})\} \tag{7.2}$$

In the above matrix equation, the values of $\{p(t_F)\}$ and $\{q(t_F)\}$ are evaluated at time t_F through a time-marching scheme, given the initial conditions at time t_{F-1}. The mass matrix cannot distinctly be identified from this time-discretized equation and hence this equation cannot be used as a basis for generalized dynamic analysis including harmonic and algebraic eigenvalue analyses. In order to formulate an algebraic eigenvalue problem, we need to resort to so-called static Green's function method outlined in Chapter 6. The foundation material for the formulation of the mass matrix is laid down in Sections 6.1.1 and 6.1.2. In this chapter, we will present the details of the formulations with illustrations. Over the last two decades or so, a number of methods of the formulation of algebraic eigenproblem have been put forth based on the static Green's function method. These are: Internal Cell Method (ICM), Dual Reciprocity Method (DRM), Particular Integral Method (PIM), and variations of the DRM and PIM. In the following section we will provide detailed description of the ICM.

7.3. Formulation of Internal Cell Method

ICM combines the boundary method (such as, BEM) and the domain method (such as, FEM) and may be considered to be the precursor to the BE algebraic eigenproblem formulations. The idea of the Static Fundamental Solution Method, presented in Chapter 6, was first introduced in the formulation of ICM. Therefore, we pick up the main idea from Chapter 6 and show the development of ICM. Let us consider once again the acoustic wave equation:

$$\nabla^2 p = \frac{1}{c^2}\frac{\partial^2 p}{\partial t^2} \tag{7.3}$$

Employing static fundamental solution to the left-hand side of this equation, we can write the boundary integrals for the left-hand term leaving the right-hand volume integral intact. This leads to equation (6.3), which is rewritten here for convenience:

$$C_p p_p + \int_\Gamma p q^* \,\mathrm{d}\Gamma - \int_\Gamma q p^* \,\mathrm{d}\Gamma + \frac{1}{c^2}\int_\Omega p^* \ddot{p} \,\mathrm{d}\Omega = 0 \tag{7.4}$$

The last integral may be identified as the inertia term, which will give us the mass matrix upon discretization. Unlike DRM, introduced in Section 6.1.1, no more integral transformation will be performed on this inertia term. Rather, the domain is here broken into a number of cells and the integration is directly performed on the volume of each cell for all the boundary nodes. We also need to consider the following boundary integral equation for the internal cells in order to arrive at the complete set of equations:

$$-p_i = \int_\Gamma p q^* \,\mathrm{d}\Gamma - \int_\Gamma q p^* \,\mathrm{d}\Gamma + \frac{1}{c^2}\int_\Omega p^* \ddot{p} \,\mathrm{d}\Omega \quad \text{(for “}i\text{”th internal cell)} \tag{7.5}$$

This is alternatively called the boundary element post-processing equation.

We illustrate the formulation using a simple example. Let us consider a rectangular acoustical cavity as shown in Figure 7.1. The boundary of the cavity is divided

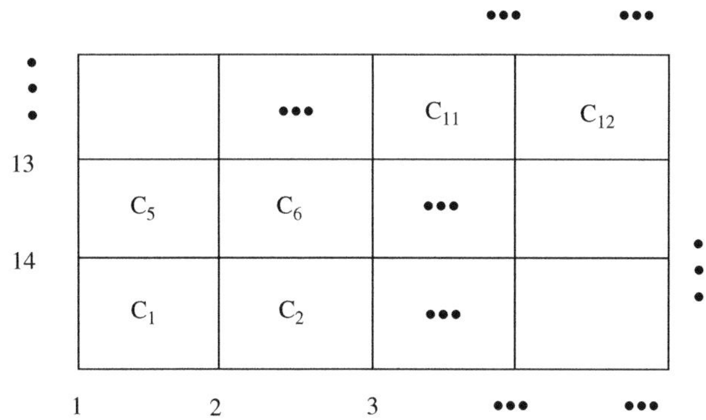

Figure 7.1. Illustration for internal cell method (ICM).

into $n = 14$ boundary element segments and the domain is broken into $m = 12$ cells. In Chapters 2 and 3 we showed the boundary discretization using polynomial shape functions of different orders to represent pressure and pressure gradients. In the same manner, the pressure within each internal cell can be interpolated using polynomial shape functions $p = N_i p_i$. Assuming shape functions to be constant, i.e., $N_i = 1$ over each segment of the boundary as well as over each internal cell, the discretized boundary integral equation (7.4) can be written as:

$$C_i p_i + \sum_{j=1}^{n} \int_{\Gamma_j} p_j q_{ij}^* \, d\Gamma_j - \sum_{j=1}^{n} \int_{\Gamma_j} q_j p_{ij}^* \, d\Gamma_j = -\frac{1}{c^2} \sum_{k=1}^{m} \int_{\Omega_k} p_{ik}^* \ddot{p}_k \, d\Omega_k \tag{7.6}$$

Here the boundary integral spans over n segments and the volume integral term spans over m internal cells ($n = 14$ and $m = 12$ in Fig. 7.1). Since there are n boundary nodes and m internal cells, the right-hand side inertia term will lead to a rectangular matrix of size $n \times m$. The matrix form of the above equation is:

$$[H]\{p\} + [G]\{q\} = -(1/c^2)[\overline{M}]\{\ddot{p}\} \tag{7.7}$$

Now the BE equation for the internal cells [eqn. (7.5)] can be discretized as follows:

$$\{\hat{p}\} = [\hat{H}]\{p\} + [\hat{G}]\{q\} + (1/c^2)[\hat{M}]\{\ddot{\hat{p}}\} \tag{7.8}$$

Equations (7.7) and (7.8) together represent the transient dynamic equation for acoustic wave propagation obtained using ICM. Note that $[H]$, $[G]$ are square matrices of size $n \times n$, $[M]$ is a rectangular matrix of size $n \times m$, $[\hat{H}]$, $[\hat{G}]$ are rectangular matrices of size $m \times n$ and $[\hat{M}]$ is a square matrix of size $m \times m$. $\{\hat{p}\}$ represents the unknown pressures within the cells in the domain.

The above equations can be combined together to set up a single set of matrix equation:

$$\begin{bmatrix} [H] & [0] \\ [\hat{H}] & -[I] \end{bmatrix} \begin{Bmatrix} p \\ \hat{p} \end{Bmatrix} + \begin{bmatrix} [G] & [0] \\ [\hat{G}] & [0] \end{bmatrix} \begin{Bmatrix} q \\ 0 \end{Bmatrix} = -\frac{1}{c^2} \begin{bmatrix} [0] & [\overline{M}] \\ [0] & [\hat{M}] \end{bmatrix} \begin{Bmatrix} \ddot{p} \\ \ddot{\hat{p}} \end{Bmatrix} \tag{7.9}$$

This is the matrix dynamic equation for the acoustic problem employing ICM. It can be used to solve the propagation of acoustic pressure wave as a function of time. The same equation can be used to set up the eigenvalue problem by the simple substitution of a time-harmonically varying pressure $p = Pe^{j\omega t}$ and $\hat{p} = \hat{P}e^{j\omega t}$. For the case of an acoustic cavity with hard boundaries the pressure gradient terms are zero, i.e., $\{Q\} = \{0\}$ and the following eigenvalue problem results:

$$\begin{bmatrix} [H] & [0] \\ [\hat{H}] & -[I] \end{bmatrix} \begin{Bmatrix} P \\ \hat{P} \end{Bmatrix} = \frac{\omega^2}{c^2} \begin{bmatrix} [0] & [\overline{M}] \\ [0] & [\hat{M}] \end{bmatrix} \begin{Bmatrix} P \\ \hat{P} \end{Bmatrix} \tag{7.10}$$

Note that both the boundary and the internal pressures are treated as unknowns here. Alternatively, the boundary pressures $\{P\}$ can be solved for from the first row of this equation:

$$\{P\} = (\omega^2/c^2)[H]^{-1}[\overline{M}]\{\hat{P}\} \tag{7.11}$$

which on substitution in the second row of equation (7.10) yields:

$$[K]\{\hat{P}\} = (\omega^2/c^2)[M]\{\hat{P}\} \tag{7.12}$$

where:

$$[K] = [I] \tag{7.13}$$

and

$$[M] = [\hat{H}][H]^{-1}[\overline{M}] - [\hat{M}] \tag{7.14}$$

The resonant frequencies and mode shapes of the cavity can be found by solving this algebraic eigenvalue problem.

7.4. Example of internal cell method: rectangular plate vibration

The ICM was developed in the context of plate vibration problems [115] in an attempt to solve for the free vibration frequencies and mode shapes of rectangular plates. The governing differential equation of thin plate vibration in terms of the transverse deflection w is given by the biharmonic equation (Kirchhoff's theory of thin plate):

$$\nabla^4 w + \frac{\rho h}{D}\frac{\partial^2 w}{\partial t^2} = 0 \quad \text{in } \Omega \tag{7.15}$$

where D = flexural rigidity = $Eh^3/12(1-\nu^2)$, h = thickness of the plate, ρ = mass density of plate material and ∇^4 = biharmonic operator = $\partial^4(\cdots)/\partial x^4 + 2\partial^4(\cdots)/\partial x^2 y^2 + \partial^4(\cdots)/\partial y^4$. For free vibration problem of plates, we can substitute the time-harmonic variation of deflection $w = We^{j\omega t}$ into the dynamic equation (7.15) to arrive at:

$$D\nabla^4 W = \rho\omega^2 W \quad \text{in } \Omega \tag{7.16}$$

It may be noted that at any given frequency of vibration ω, this equation represents the static plate deflection problem under uniform pressure loading whose value is the same as the inertia loading ($\rho h\omega^2 W$). We can therefore utilize a fundamental solution corresponding to the static plate vibration problem to formulate the problem:

$$D\nabla^4 W^*(p,q) = \delta(p,q) \tag{7.17}$$

This is nothing but the static Green's function method used above in the context of acoustics eigenvalue formulation. The alternative is to use fundamental solution corresponding to the equation $D\nabla^4 W - \rho h\omega^2 W = 0$. Bezine [115] felt that the numerical integrals with this fundamental solution, which involves Hankel function, is awkward and therefore he proposed the static Green's function method. However, we note that this method not only avoided the use of complicated fundamental solutions, but led to an algebraic eigenvalue formulation also. As we already know, this fundamental solution implicitly contains the frequency of vibration ω in it and so it prevents the direct formulation of an algebraic eigenvalue problem.

The static Green's function or the fundamental solution W^* for the unit point load $\delta(p,q)$ of equation (7.17) is found to be:

$$W^* = \frac{r^2(p,q)}{8\pi D} \ln\{r(p,q)\} \tag{7.18}$$

As in the case of acoustics, we can proceed to write BE equation for the left-hand side of equation (7.16) keeping the right-hand side inertia term intact and using the above fundamental solution:

$$\begin{aligned} C(P)W(P) &+ \frac{1}{D}\int_\Gamma K(W^*(P,Q))W(Q)\,\mathrm{d}\Gamma - \frac{1}{D}\int_\Gamma M(W^*(P,Q))\frac{\partial W(Q)}{\partial n_Q}\,\mathrm{d}\Gamma \\ &+ \frac{1}{D}\int_\Gamma \frac{\partial W^*(P,Q)}{\partial n_Q} M(W(Q))\,\mathrm{d}\Gamma - \frac{1}{D}\int_\Gamma W^*(P,Q)K(W(Q))\,\mathrm{d}\Gamma \\ &+ \frac{1}{D}\sum_{k=1}^{N}[W(A_k)T(W^*(P,A_k)) - T(W(A_k))W^*(P,A_k)] \\ &= \frac{\rho h}{D}\omega^2 \int_\Omega W^*(P,Q)W(Q)\,\mathrm{d}\Omega \end{aligned} \tag{7.19}$$

where:

P and Q are respectively the source and field points on the boundary Γ;
n_Q and n_P are respectively the outward normal at the points Q and P;
$K(W^*(P,Q))$ is the Kirchhoff transverse shear force corresponding to the deflection $W^*(P,Q)$;
$M(W^*(P,Q))$ is the normal flexural moment corresponding to the deflection $W^*(P,Q)$;
$T(W^*(P,Q))$ is the torsional moment corresponding to the deflection $W^*(P,Q)$; and
$[.]_{A_k}$ is the jump of the function which may occur at corners A_k of curvilinear abscissa s_k, defined by $[.]_{A_k} = (.)^+_{s_k} - (.)^-_{s_k}$.

Also,

$$T(W) = -D(1-\nu)\frac{\partial}{\partial s}\left(\frac{\partial W(Q)}{\partial n_Q}\right)$$

Unlike acoustic BE formulation, the plate vibration problem will require an additional set of equations since we have an additional pair of unknowns per node M and K [71].

This second equation is obtained by differentiating equation (7.19) with respect to the normal:

$$C(P)\frac{\partial W(P)}{\partial n_P} + \frac{1}{D}\int_\Gamma \frac{\partial K(W^*(P,Q))}{\partial n_P} W(Q)\,\mathrm{d}\Gamma - \frac{1}{D}\int_\Gamma \frac{\partial M(W^*(P,Q))}{\partial n_P}\frac{\partial W(Q)}{\partial n_Q}\,\mathrm{d}\Gamma$$
$$+\frac{1}{D}\int_\Gamma \frac{\partial^2 W^*(P,Q)}{\partial n_P \partial n_Q} M(W(Q))\,\mathrm{d}\Gamma - \frac{1}{D}\int_\Gamma \frac{\partial W^*(P,Q)}{\partial n_P} K(W(Q))\,\mathrm{d}\Gamma$$
$$+\frac{1}{D}\sum_{k=1}^{N}\left[W(A_k)\frac{\partial T(W^*(P,A_k))}{\partial n_P} - T(W(A_k))\frac{\partial W^*(P,A_k)}{\partial n_P}\right]$$
$$= \frac{\rho h}{D}\omega^2 \int_\Omega \frac{\partial W^*(P,Q)}{\partial n_P} W(Q)\,\mathrm{d}\Omega \tag{7.20}$$

The homogeneous boundary conditions for the plate can be one of the following:

$$\begin{aligned} &\text{Simple support:} \quad W = M(W) = 0 \\ &\text{Clamped:} \quad W = \frac{\partial W}{\partial n_Q} = 0 \\ &\text{Free edge:} \quad K(W) = M(W) = 0 \end{aligned} \tag{7.21}$$

As in the case of acoustics described in the last section, we can divide the boundary of the plate Γ into "n" boundary element segments and the plate domain Ω into "m" internal cells. If we use constant elements, then each BE segment will have four unknowns: W, $\partial W/\partial n_Q$, M and K and each internal cell will have one unknown, which is the deflection itself W. The BE equations (7.19) and (7.20) can now be integrated over each BE segment and each internal cell to yield:

$$[A_\Gamma]\{Y_\Gamma\} - \omega^2[M_\Gamma]\{W_\Omega\} = \{0\} \tag{7.22}$$

This matrix equation is a discretized form of both boundary integral equations (7.19) and (7.20). Also, we already applied the applicable homogeneous boundary conditions from (7.21). Thus:

$[A_\Gamma]$ is a $2n \times 2n$ matrix resulting from the line integrals of equations (7.19) and (7.20);
$\{Y_\Gamma\}$ is a vector of $2n$ remaining unknowns on the boundary Γ;
$[M_\Gamma]$ is a $2n \times m$ matrix obtained by performing the domain integrals in equations (7.19) and (7.20);
$\{W_\Omega\}$ is a vector of m unknown deflections in the internal cells of the plate domain Ω.

Similar to the acoustics eigenvalue formulation, we can write the BE equations for the internal cells of the plate domain:

$$\{W_\Omega\} = [A_\Omega]\{Y_\Gamma\} - \omega^2[M_\Omega]\{W_\Omega\} \tag{7.23}$$

Here $[A_\Omega]$ is a $m \times 2n$ rectangular matrix and $[M_\Omega]$ is a $m \times m$ matrix. We can now combine matrix equations (7.22) and (7.23) into a single matrix equation:

$$\begin{bmatrix} [A_\Gamma] & [0] \\ [A_\Omega] & -[I] \end{bmatrix}\begin{Bmatrix} Y_\Gamma \\ W_\Omega \end{Bmatrix} - \omega^2\begin{bmatrix} [0] & [M_\Gamma] \\ [0] & [M_\Omega] \end{bmatrix}\begin{Bmatrix} Y_\Gamma \\ W_\Omega \end{Bmatrix} = \begin{Bmatrix} 0 \\ 0 \end{Bmatrix} \tag{7.24}$$

This represents a generalized eigenvalue problem and can be solved to compute the eigenvalues ω^2 and the eigenvectors comprising the unknown degrees of freedom $[Y_\Gamma \ W_\Omega]$. Or, one can eliminate the unknowns $\{Y_\Gamma\}$ between the equations (7.22) and (7.23) in favor of the deflection degrees of freedom at the internal cells $\{W_\Omega\}$.

Let us consider a square plate made of steel with Poisson's ratio 0.3. The boundary of the plate is divided into 48 BE segments and the domain is divided into (a) $4 \times 4 = 16$, or (b) $8 \times 8 = 64$ internal cells (Fig. 7.2). The frequency of vibrations is computed in a non-dimensional form. The eigenvalue solutions for plates with all three boundary conditions noted in equations (7.21) are illustrated.

(a) For the cantilever plate, the first five frequencies are presented in Table 7.1 where analytical frequency solutions obtained by Ritz method [174] are also shown. The results from the ICM are within 3% of the Ritz method solution for the first five frequencies. The mode shapes for the first five modes computed by ICM are shown in Figures 7.3 through 7.7.

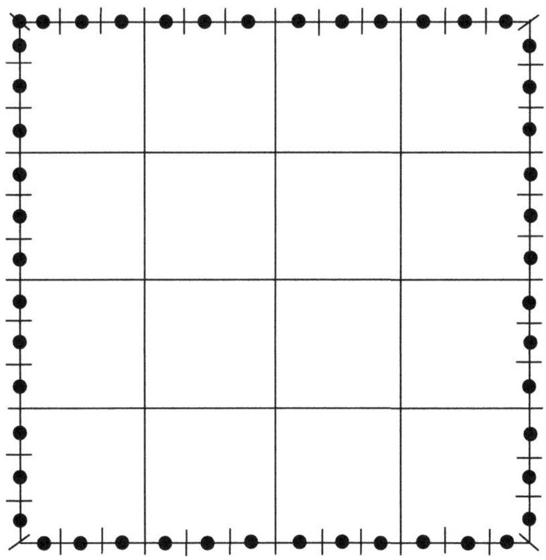

Figure 7.2. BE discretization of square plate for ICM.

Table 7.1. Frequency of vibrations for cantilever plate using ICM.

	Internal cell BE method			
Modes	4×4 cells	8×8 cells	Error (%)	Ritz method
1	3.517	3.484	0.3	3.494
2	8.805	8.571	0.3	8.547
3	24.488	22.525	0.5	21.44
4	30.879	28.104	2.4	27.46
5	33.537	31.359	0.6	31.17

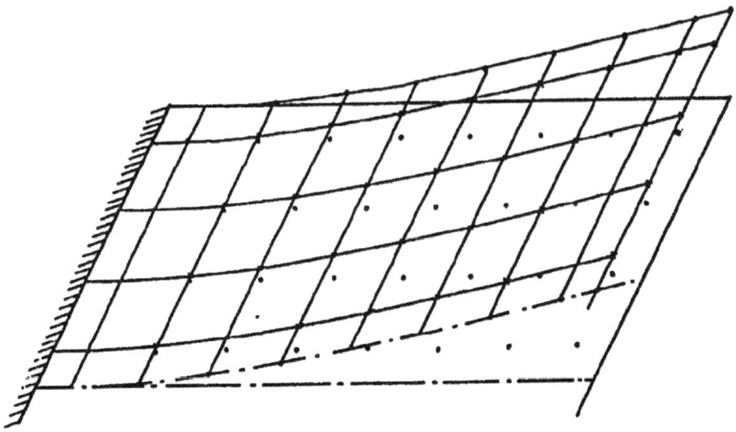

Figure 7.3. First mode of vibration for the cantilever square plate. Reproduced from ref. [115] with permission from the publisher, Elsevier Science Publishers.

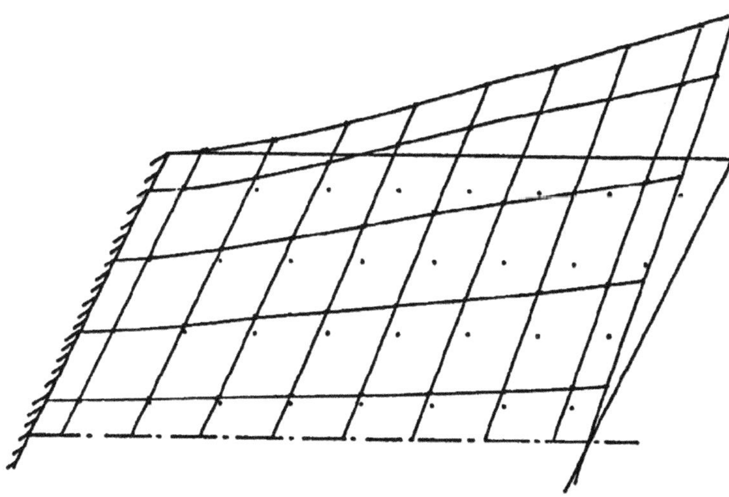

Figure 7.4. Second mode of vibration for the cantilever square plate. Reproduced from ref. [115] with permission from the publisher, Elsevier Science Publishers.

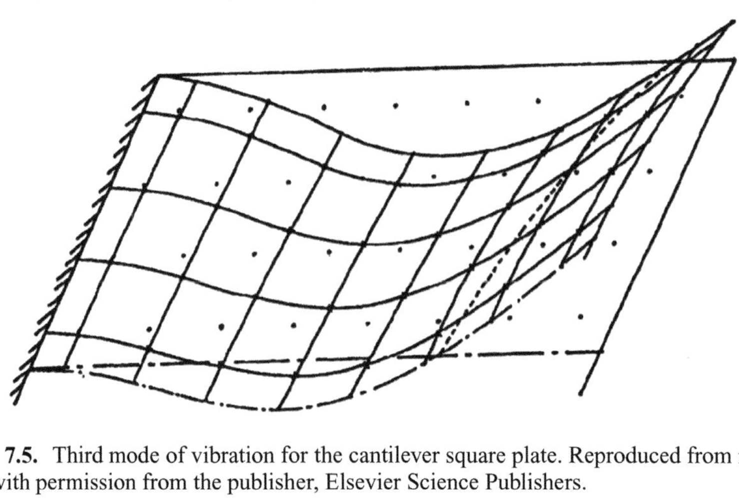

Figure 7.5. Third mode of vibration for the cantilever square plate. Reproduced from ref. [115] with permission from the publisher, Elsevier Science Publishers.

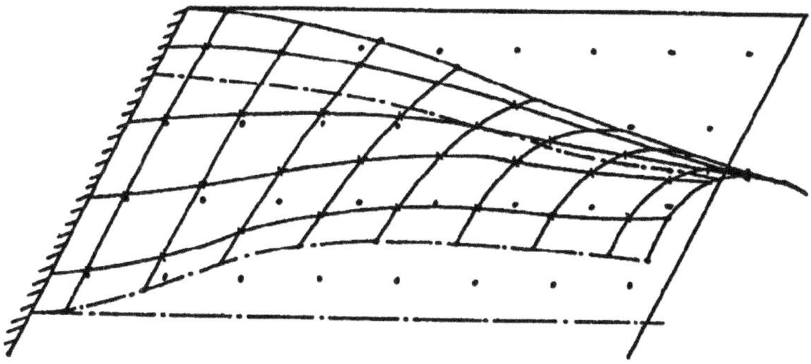

Figure 7.6. Fourth mode of vibration for the cantilever square plate. Reproduced from ref. [115] with permission from the publisher, Elsevier Science Publishers.

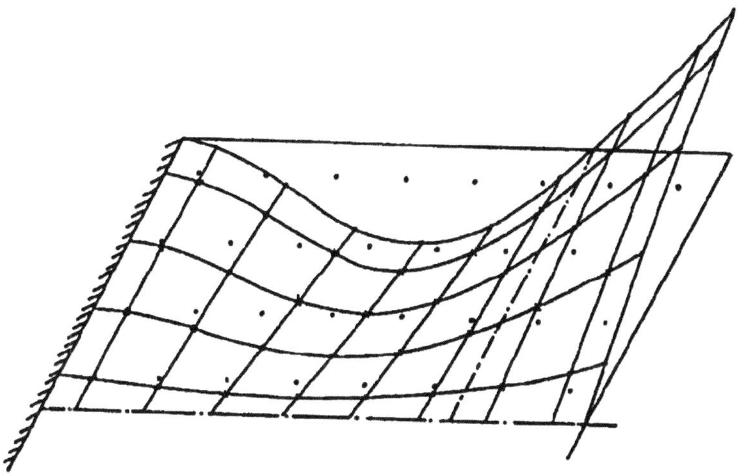

Figure 7.7. Fifth mode of vibration for the cantilever square plate. Reproduced from ref. [115] with permission from the publisher, Elsevier Science Publishers.

Table 7.2. Frequency of vibrations for clamped plate using ICM.

	Internal cell BE method				
Modes	4 × 4 cells	Error (%)	8 × 8 cells	Error (%)	Ritz method
1	37.157	3.2	36.24	0.7	35.99
2	79.721	8.6	74.77	1.8	73.41
3	122.78	13.4	111.3	2.8	108.3
4	164.07	25	136.3	3.6	131.6
5	169.05	28	137	3.6	132.3
6	201.03	22	172.7	4.6	165.1

(b) The frequency of vibration results for the clamped plate is presented in Table 7.2. The first six frequencies are recorded in the table along with the Ritz method [174] solutions for comparison. The first frequency obtained using ICM is seen to be within 1% of the Ritz method solution. For the 8 × 8 internal cell grid, the sixth frequency from ICM is within 5% compared to that of reference [174].

Table 7.3. Frequency of vibrations for simply supported plate using ICM.

	Internal cell BE method				
Modes	4×4 cells	Error (%)	8×8 cells	Error (%)	Exact solution
1	20.252	2.6	19.866	0.6	19.74
2	52.491	6.4	50.145	1.6	49.34
3	86.565	9.6	80.971	2.5	78.96
4	107.904	9.4	101.864	3.2	98.69
5	142.15	10.8	133.681	4.2	128.3

(c) A closed-form analytical solution for the frequencies of vibration of a square plate with simply supported boundary conditions is available in the textbook by Timoshenko and Woinowsky-Krieger [175]. The first five frequencies obtained using ICM are reported in Table 7.3 along the closed-form solutions. The first frequency is within 1% of the exact solution for the 8×8 grid internal cells. The fifth frequency is about 4% apart.

Chapter 8

Algebraic Eigenvalue Problem in Boundary Elements

8.1. Introduction

In the previous chapter we used the idea of a static fundamental solution method and demonstrated its application to a mixed boundary and domain technique, which we designated as the Internal Cell Method (ICM). The boundary integral technique was applied to the given problem but the inertia effects were excluded in the boundary element formulation. The domain was later divided into a grid of internal cells in order to compute the inertia term. The approach of using static fundamental solution, free of the frequency parameter, for the dynamic problem and the separate calculation of the inertia term for computing the mass or mass-type matrix laid the foundation for the boundary element algebraic eigenvalue formulation. However, the ICM required the discretization of the domain as well. This somewhat destroys the advantages of the boundary element method (BEM), which is supposed to be a boundary-only method, where the discretization is confined to the boundary alone.

This chapter will present two methods, which will subject the inertia term in equations (7.9) and (7.10) to further transformation leading to boundary-only formulation. These boundary element algebraic eigenvalue formulation methods are known as the Dual Reciprocity Method (DRM) and Particular Integral Method (PIM). The primary motivation for these methods comes from the use of static fundamental solution as in ICM described in the previous chapter. The methods then take the next step of removing the restriction of having to discretize the domain.

We will develop the BE algebraic eigenvalue formulation for acoustics using DRM in the following two sections. The development of BE algebraic eigenvalue formulation for elasticity using PIM will be shown in Sections 8.4 and 8.5. The relative merits of these two methods will also be outlined.

8.2. Eigenproblem using dual reciprocity method in acoustics

As pointed out above, the idea of the development of a separate mass-type matrix in the context of BEM was first introduced in the ICM. However, this approach was formalized by Nardini and Brebbia [116] who published the idea of separating the free vibration differential equation into two components, one, free of the frequency parameter, leading to the stiffness-type matrix and the other, inertial term containing the frequency parameter in it, leading to the system mass matrix. The former was transformed into the boundary integral equation with the use of a fundamental solution, independent of the frequency. On the other hand, the latter became a volume integral just like a body force term in the BEM. The dependent variable, pressure in the volume integral (displacement components for the eigenvalue problem in elasticity), was expressed in terms of a global shape function and fictitious density function. The volume integral was then transformed into boundary integrals with the help of Gauss's divergence theorem. The free vibration problem could, then, be cast into a generalized eigensystem. Nardini and Brebbia [120] later called this technique the "DRM".

The DRM is based on the Static Green's Function Method presented in Section 6.1.1. In the next section we shall show how the Static Green's Function Method extends to DRM which leads to the algebraic eigenvalue problem in BEM. Details on the choice of global shape functions that help in transforming volume integrals to the boundary will also be shown. A number of example acoustic eigenvalue problems will be presented in Section 8.4.

8.2.1. Development of algebraic eigenvalue problem (DRM)

Here again, as in Chapter 6, we start with the acoustic Helmholtz equation. The harmonic pressure variation of a compressible fluid is governed by:

$$\nabla^2 P + k^2 P = 0 \tag{8.1}$$

subject to the Dirichlet and/or Neumann boundary conditions:

$$P = 0 \quad \text{on open surface } \Gamma_1 \tag{8.2}$$

$$Q = \partial P/\partial n = 0 \quad \text{on acoustically hard surface } \Gamma_2 \tag{8.3}$$

respectively. Or, in case the fluid domain is bounded by a harmonically vibrating structure, the Neumann boundary condition takes the form:

$$Q = \partial P/\partial n = \rho\omega^2 u_n \tag{8.4}$$

In equations (8.1) through (8.4),

P is the amplitude of the pressure;
k is the wave number $= \omega/c$;
ω is the circular frequency in radians/second;
c is the speed of sound through the fluid medium;
ρ is the density of the fluid, and
u_n is the normal component of the displacement of the structure at fluid–structure interface.

The problem can be defined as follows: given an enclosure filled with a fluid, e.g., air, having acoustically hard ($Q = \partial P/\partial n = 0$) and partly open ($P = 0$) boundaries, how to find the eigenvalues (ω) and eigenmodes (P) of the acoustic fluid within the enclosure. The case when the fluid is coupled to a harmonically vibrating structure (boundary condition represented by equation 8.4) will be discussed in Chapter 10. In this chapter we will only deal with the most common boundary condition represented by equation (8.3).

Note that we dealt with the original governing dynamic equation of motion in Chapter 6 in deriving the general discretized matrix equation of motion using the so-called Static Green's Function Method (eqn. 6.19). In that formulation, we can now assume harmonic oscillation of the acoustic pressure, substitute $p = Pe^{j\omega t}$ in equation (6.19) and set up the algebraic eigenvalue problem. However, here we chose to make the assumption of harmonic pressure variation from the outset. The formulation will then directly lead to an algebraic eigenvalue problem.

The boundary integral equation corresponding to equation (8.1) with the pressure amplitude $P(x)$ as the dependent variable is written as:

$$C_P P(\xi) + \int_\Gamma P(x) Q^*(x,\xi)\,\mathrm{d}\Gamma(x) - \int_\Gamma P^*(x,\xi) Q(x)\,\mathrm{d}\Gamma(x) + k^2 \int_\Omega P^*(x,\xi) P(x)\,\mathrm{d}\Omega = 0 \tag{8.5}$$

where:

C_P is a geometric coefficient at the source point ξ;
n is the outward normal at the field point x;
Ω is the domain and Γ is the boundary of the domain;
$P^*(x,\xi)$ is the fundamental solution to the Laplace's equation given by equations (2.8) or (2.9);
r is the distance between x and ξ.

Note that this BE equation is same as equation (6.3), the only difference being that here we are dealing with the pressure amplitude $P(x)$ and not the time-dependent pressure $p(x,t)$ as the dependent variable. In equation (8.5) the inertia term involving the frequency parameter k was not transformed. The full treatment of the Helmholtz equation (8.1) would call for a different fundamental solution, P^*, involving a Hankel function in two-dimensional (2-D) and an exponential function in three-dimensional (3-D). After discretization of the boundary Γ, the equation (8.5) can be written as:

$$[G]\{Q\} - [H]\{P\} = k^2 \int_\Omega P^*(x,\xi) P(x)\,\mathrm{d}\Omega \tag{8.6}$$

Now, if it is possible to find a function Ψ such that:

$$\nabla^2 \Psi = P(x) \tag{8.7}$$

inside the domain, equation (8.6), with the application of Gauss's divergence theorem to the right-hand side, can be written in terms of boundary integrals alone:

$$[G]\{Q\} - [H]\{P\} = k^2(-[G]\{\partial\Psi/\partial n\} + [H]\{\Psi\}) \tag{8.8}$$

A global shape function can now be introduced in order to approximate the pressure inside the domain:

$$P(x) = \sum_{m=1}^{\infty} C(x,\xi_m)\Phi(\xi_m) \tag{8.9}$$

where:

x is a point in the domain;
ξ_m is a source point at the boundary; and
Φ is a fictitious density function at ξ_m.

The most widely used global shape function is given as:

$$C(x, \xi_m) = \mathrm{R} - r(x, \xi_m) \tag{8.10}$$

where R is a suitable constant, e.g., the largest distance between any two points in the body. Other forms of shape functions that can be used will be discussed in next chapter. This shape function can be inserted into equation (8.9) to express the pressure amplitude $P(x)$ in terms of the shape functions. This pressure amplitude can then be substituted into equation (8.7) to yield:

$$\nabla^2 \Psi = \sum_{m=1}^{\infty} \{\mathrm{R} - r(x, \xi_m)\} \Phi(\xi_m) \tag{8.11}$$

This differential equation can be solved for Ψ:

$$\Psi = -\sum_{m=1}^{\infty} D(x, \xi_m) \Phi(\xi_m) \tag{8.12}$$

where:

$$D(x, \xi_m) = r^3/d_1 - \mathrm{R}r^2/d_2 \tag{8.13}$$

$d_1 = 3(d+1)$ and $d_2 = 2d$; $d = 2$ and 3 for 2-D and 3-D problems, respectively. Equation (8.8) also contains the normal derivative of the function Ψ at point x, which can be computed from equation (8.12) as:

$$\frac{\partial \Psi}{\partial n} = -\sum_{m=1}^{\infty} T(x, \xi_m)\ \Phi(\xi_m) \tag{8.14}$$

where:

$$T(x, \xi_m) = (3r^2/d_1 - 2\mathrm{R}r/d_2)\ \partial r/\partial n \tag{8.15}$$

Let us now substitute the expressions of the function Ψ and its normal derivative $\partial \Psi/\partial n$ into equation (8.8) and obtain:

$$[G]\{Q\} - [H]\{P\} = k^2(-[G][T] + [H][D])\{\Phi\} \tag{8.16}$$

This equation contains both the physical pressure P and the fictitious density function Φ as the dependent variable. We can eliminate the fictitious function Φ and cast this equation entirely in terms of pressure amplitude P and its normal derivative $\partial P/\partial n$. To this end, we can use the discretized boundary nodal points as the only collocation points and write the equation (8.9) in a matrix form:

$$\{P\} = [C]\{\Phi\} \tag{8.17}$$

We can solve for the fictitious density function from this equation as:

$$\{\Phi\} = [C]^{-1}\{P\} \tag{8.18}$$

Using this, equation (8.16) can be written entirely in terms of pressure amplitude and its normal derivative:

$$[G]\{Q\} - [H]\{P\} = k^2([G][T] - [H][D])[C]^{-1}\{P\} \tag{8.19}$$

Let us write this equation as:

$$[G]\{Q\} - [H]\{P\} = k^2[M]\{P\} \tag{8.20}$$

where $[M]$ may be recognized as the mass-type matrix, given by:

$$[M] = ([G][T] - [H][D])[C]^{-1} \tag{8.21}$$

After applying appropriate boundary conditions, given in equations (8.2) and (8.3), equation (8.20) can be put in the form of generalized eigenproblem:

$$[A]\{x_i\} = k^2[B]\{x_i\} \tag{8.22}$$

where:

$\{x_i\}$ are the eigenvectors $\lfloor P_i, Q_i \rfloor$
k_i are the eigenvalues ($k_i = \omega_i/c$)

This method requires the inversion of a matrix [eqn. (8.18)] so as to formulate the algebraic eigenvalue problem in terms of the physical pressure. It has its advantages since both the types of boundary conditions represented by equations (8.2) and (8.3) can easily be handled. Chapter 9 will present a modified version of DRM/PIM in which the inversion of the matrix will be avoided [131].

8.2.2. Example problems of acoustic eigenvalue analysis (DRM)

A number of examples are presented in this section in order to demonstrate the validity and accuracy of the DRM. The boundary is discretized by linear and quadratic elements. The boundary walls for the problems are considered acoustically hard ($Q = \partial P/\partial n = 0$). The Lanczos eigensolver, discussed in Chapter 11, is employed to extract the eigenvalues.

Example 8.1: Impedance tube

The impedance tube problem is suitable for validating acoustic eigenvalue solution methods. The boundary walls of the impedance tube is assumed to be acoustically hard, $\partial P/\partial n = 0$. The speed of sound "$c$" in air is taken as 340 meter/second. When the length of the tube "a" is much greater than the width "b", i.e., for $a \gg b$ and for the lower mode shapes, the acoustic behavior of the tube may be considered as one-dimensional, for which a closed form analytical solution for the resonant frequency in Hertz is available:

$$f = cn/(2a) \quad \text{where } n = 1, 2, 3, \ldots, (a/b) \tag{8.23}$$

A convergence study is performed for this problem for the first five modes and the results are shown in Figure 8.1. The short side of the tube with length "b" is always modeled with one quadratic element. The longer side with length "a" is divided into 1, 2, 4, 6, 8, 10 and 12 boundary element segments. The solutions are seen to converge rapidly. The second mode, for example, converges to within 2% when four elements are used in the longitudinal direction. The fifth mode is seen to yield the resonant frequency within 2% when only eight elements are used along the length of the tube.

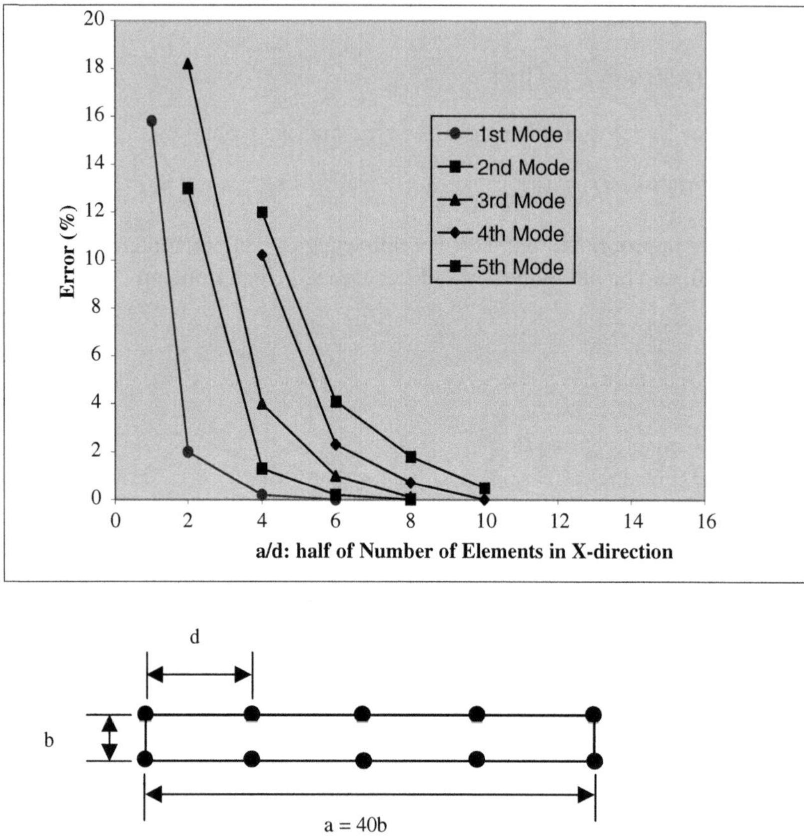

Figure 8.1. Convergence study of impedance tube resonant frequencies.

Example 8.2: Automotive passenger cabin without seats

The acoustical characteristics of the automotive passenger cabin can be conveniently investigated using a boundary element eigenvalue method such as DRM. The computation of resonant frequencies of such passenger cabins is of importance as these cabins must be designed to have resonant frequencies away from the frequencies of the vibrating components of the operating automobile, such as, the engine or the car stereo. In case the cabin frequency resonates with that of the vibrating automotive components, it can cause great discomfort to the passengers.

Here we discuss the application of the BE eigenvalue method to the automotive cabin enclosure. Figure 8.2 shows a simple 2-D BE model of a small passenger car cabin without the seats. The entire cabin is modeled using 23 quadratic boundary elements. The walls of the cabin are assumed to be acoustically hard surfaces. The speed of sound in the enclosure air is taken as 340 meter/second. This particular problem was also solved by Shuku and Ishihara [176]. They used finite element method (FEM) and also conducted an experimental study on the passenger compartment without the seats. The results of their studies along with the boundary element eigensolutions are presented in Table 8.1 for the first three resonant frequencies. The finite element, boundary element and experimental solutions are seen to agree well.

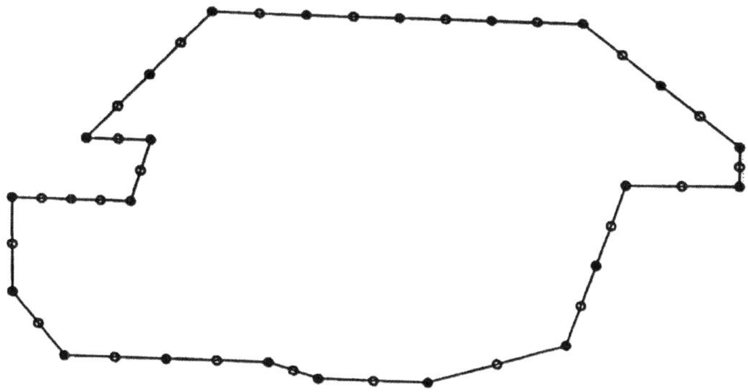

Figure 8.2. Two-dimensional boundary element model of a car passenger cabin without seats. Reproduced from ref. [124] with permission from the publisher, John Wiley & Sons.

Table 8.1. Resonant frequencies of automotive cabin: FEM, BEM and experimental results.

Mode number	Experimental results	FEM solution	BEM solution
1	87.5	86.8	87.6
2	138.5	138.0	138.7
3	157.0	154.6	153.2

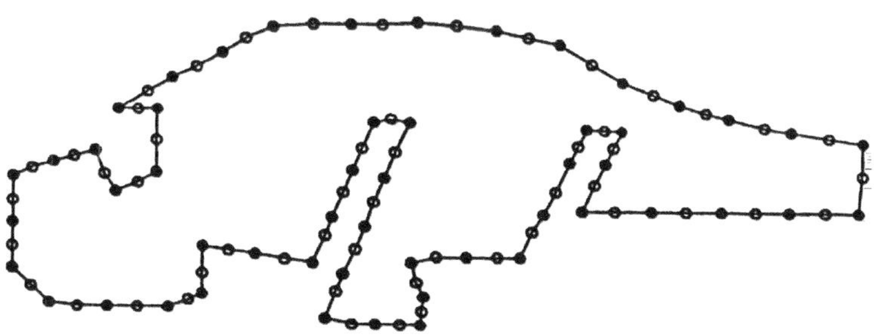

Figure 8.3. One-region BE model of a hatchback automobile cabin with seats. (•) End node; (o) midnode. Reproduced from ref. [124] with permission from the publisher, John Wiley & Sons.

Example 8.3: Automotive passenger cabin: effects of including seats

The inclusion of the car seats in the model may drastically alter the acoustic character of the automobile compartment. A two-dimensional (2-D) model of the cabin enclosure of a hatchback automobile is considered in this example. Four different cases are identified:

(a) Use one boundary element region and include seats;
(b) Use four boundary element regions and include seats;
(c) Use one boundary element region and do not include seats;
(d) Use four boundary element regions and do not include seats.

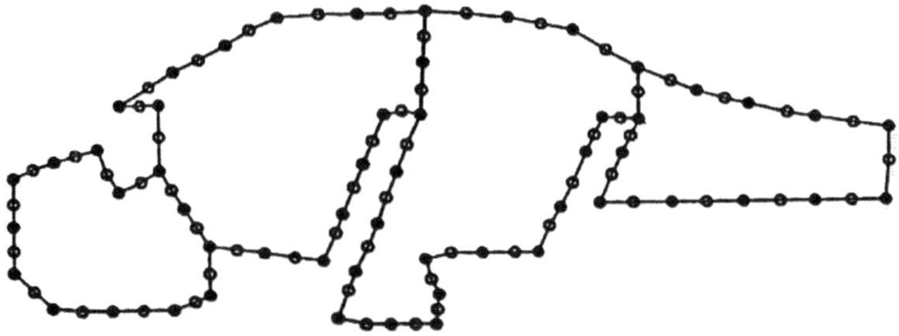

Figure 8.4. Four-region BE model of a hatchback automobile cabin with seats. (•) End node; (○) midnode. Reproduced from ref. [124] with permission from the publisher, John Wiley & Sons.

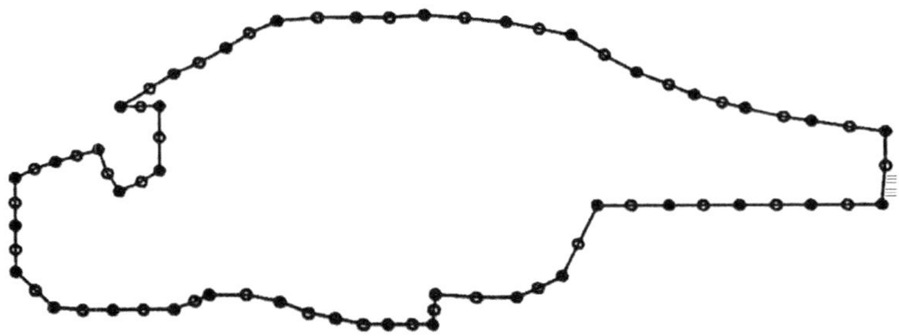

Figure 8.5. One-region BE model of a hatchback automobile cabin without seats. (•) End node; (○) midnode. Reproduced from ref. [124] with permission from the publisher, John Wiley & Sons.

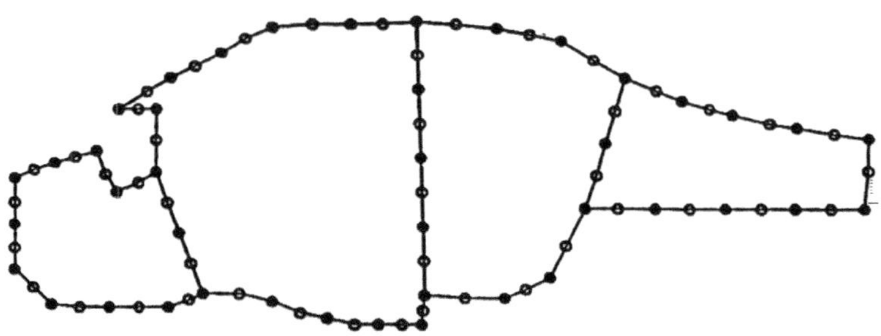

Figure 8.6. Four-region BE model of a hatchback automobile cabin without seats. (•) End node; (○) midnode. Reproduced from ref. [124] with permission from the publisher, John Wiley & Sons.

In all cases, the walls of the automobile compartment are considered acoustically hard surfaces. The speed of sound through the air in the compartment is once again assumed to be 340 m/s and three-noded quadratic boundary elements are used. The one-region BE model of the car with the seats included is shown in Figure 8.3, whereas the four-region BE model with the seats is shown in Figure 8.4. Also, the one-region and four-region BE models of the car cabin without the seats are represented in Figures 8.5 and 8.6, respectively.

Table 8.2. Resonant frequencies of a hatchback automotive cabin: FEM, BEM and experimental results.

	Without seats				With seats			
			BEM				BEM	
Mode	Experiment	FEM	Mesh 1	Mesh 2	Experiment	FEM	Mesh 1	Mesh 2
1	60	68	69	72	53	50	49	49
2	110	105	104	110	–	79	75	78
3	135	152	153	155	–	125	116	122
4	–	179	190	186	–	163	193	159

Mesh 1: One-region BE mesh.
Mesh 2: Four-region BE mesh.

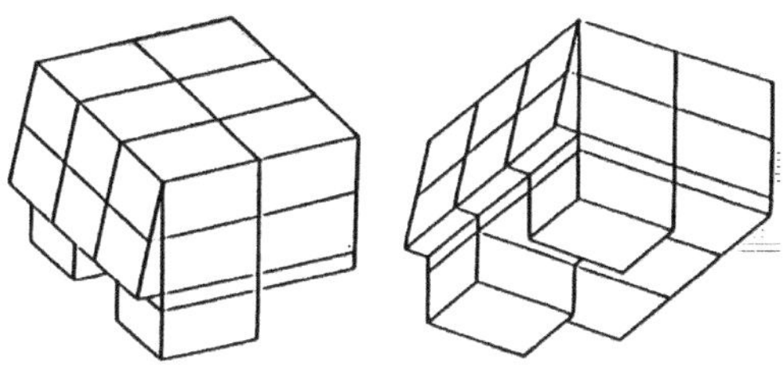

Figure 8.7. Three-dimensional BE model of a truck cab. Reproduced from ref. [124] with permission from the publisher, John Wiley & Sons.

These acoustic eigenvalue problems were also studied by Nefske, Wolf and Howell [177] who analyzed the problem using 2-D FEM and compared their solutions with experimental results. This study did not report the exact dimensions of the automobile compartment. The BE model is built from actual measurements of a hatchback car compartment which closely resembles the model of reference [177]. The BEM, FEM and experimental results are compared against each other for all the cases and are shown in Table 8.2. According to Nefske, Wolf and Howell, the discrepancy between their FEM and experimental solutions are due to the contribution of the structural flexibility of the enclosing walls, i.e., in actuality, the walls were not fully acoustically hard as it was assumed in the FEM and BEM analyses. Even though the BE analysis is performed on approximated geometric data of the cabin, the BEM solutions are seen to be remarkably close to the results presented in reference [177].

Example 8.4: Truck cab in three dimensions
An example of a 3-D BE eigenvalue analysis is now presented. A truck cab approximately 7 ft $\times$ 6 ft $\times$ 5 ft 6 inches in overall dimensions was modeled using 55 eight-noded quadratic serendipity quadrilateral boundary elements and two six-noded quadratic triangular boundary elements (Fig. 8.7). The total number of boundary element nodes for this discretization turns out to be 171. The boundary walls once again are considered as acoustically hard surfaces and the speed of sound is taken as 340 m/s. This problem was also investigated by Nefske, Wolf and Howell [177] in which they used 3-D finite elements. Here also they did not report any dimension of the truck cab. The above dimensions closely resembled the model used in reference [177] and

Table 8.3. Resonant frequencies of a truck cab by FEM and BEM.

Mode number	FEM solution	BEM solution
1	67	72
2	82	85
3	105	109

are thought to be appropriate for a standard truck cab. The BEM and FEM results are presented in Table 8.3. It is seen that despite the uncertainty of the geometric dimensions of the truck cab between the analyses the results are remarkably close to each other.

The DRM, as presented above, is certainly elegant because it preserves the boundary-only nature of the BEM and it leads to an algebraic eigenvalue formulation. However, DRM or PIM, as applied to the acoustic eigenvalue analysis, will yield poor answers or at times may miss an eigenfrequency in the complete solution. It will be shown in Chapter 9 that in these cases the breakup of the domain into multiple regions is not only a convenience but a necessity. Another alternative to achieving higher accuracy in computing resonant frequencies by DRM (or PIM) is to insert internal collocation points into the domain. The mechanism of putting internal collocation points into the domain will be discussed in Chapter 9. Also, the use of alternative global shape functions [eqn. (8.10)] for approximating the pressure inside the domain by the boundary collocation points will be investigated in Chapter 9.

Pursuing on the same lines as DRM, but by taking a slightly different mathematical route, the PIM also leads to the same algebraic eigenvalue problem as in equation (8.22). We shall describe in detail the PIM approach by considering the free vibration response problem in elasticity in the following section.

8.3. Eigenproblem using particular integral method in elasticity

The PIM is an alternative technique to transform domain integrals into boundary-only integrals and can be employed to acoustic eigenvalue problems as well as free-vibration problems in elasticity. This method was originally proposed by Ahmad and Banerjee [123], who applied the method to formulate boundary element algebraic eigenvalue problems. They developed the method for the free-vibration problems in 2-D and 3-D elasticity and applied it to solve a number of 2-D problems. Since then Banerjee and his co-workers extended the application of PIM to the cases of acoustic eigenvalue analysis [124, 134], free vibration analysis of 3-D and axisymmetric solids [125, 127] and non-axisymmetric free vibration analysis of axisymmetric solids [126]. They also applied PIM to other areas, such as, thermo-elasticity, elasto-dynamics, etc., which are beyond the scope of this book.

Similar to the DRM, the PIM treats the inertia term of the differential equation separately. Here the total solution variable is looked upon as being composed of two distinct parts: the complementary functions and the particular integrals. The boundary element integral is formulated as usual for the complementary function which does not include the frequency parameter in it. The original differential equation is posed again with the particular integral part of the solution, treating the inertia term as the forcing function. As in the DRM, the total solution variable appearing in the inertia term is approximated using global shape functions and the particular integral is solved for. The complementary function and the particular integral are added together to obtain the final solution. In the process, a distinct mass-type matrix is formed which puts the frequency parameter outside the matrix as a multiplier.

As applied to structural free-vibration problems, PIM (and DRM) will produce vibration frequency solutions that are dependent on the level of discretization of the boundary only and will not require internal collocation points for improving solution accuracy. However, when PIM is applied to acoustic algebraic eigenvalue analysis [124], the global shape function used to approximate the total pressure in the inertia term is not adequate to yield accurate solutions for many situations. In these cases, it will be necessary to insert internal collocation points to obtain more accurate resonant frequencies.

We will present the detailed description of the PIM, as applied to free-vibration problems, in the following section. We will develop formulations for 2-D and 3-D cases.

8.3.1. Development of algebraic eigenvalue problem (PIM)

The time-harmonic governing differential equation of an elastic, homogeneous and isotropic body can be written as:

$$\mu u_{i,jj} + (\lambda + \mu)u_{j,ji} + \rho\omega^2 u_i = 0 \tag{8.24}$$

This is the elastodynamic equilibrium equation for harmonic vibration written in terms of displacements in the absence of body forces. In the above equation, μ is the shear modulus (eqn. 6.21), λ is the Lamé's constant and ρ is the density. For convenience, this equation will be written in terms of a differential operator $\mathcal{L}$, which can be defined as:

$$\mathcal{L}(\cdots) = \left[\mu\,(\cdots)_{jj} + (\lambda + \mu)\,(\cdots)_{ji}\right] \tag{8.25}$$

As a result, equation (8.24) can be written in a compact form:

$$\mathcal{L}(u_i) + \rho\omega^2 u_i = 0 \tag{8.26}$$

In PIM, the total displacement solutions u_i and the tractions t_i on the boundary are divided into complementary functions and particular solutions in the following manner:

$$u_i = u_i^{\mathrm{c}} + u_i^{\mathrm{p}} \tag{8.27}$$

$$t_i = t_i^{\mathrm{c}} + t_i^{\mathrm{p}} \tag{8.28}$$

The complementary function, u_i^{c}, satisfies the first term of equation (8.26):

$$\mathcal{L}(u_i^{\mathrm{c}}) = 0 \tag{8.29}$$

The particular solution, u_i^{p}, on the other hand, satisfies the following equation:

$$\mathcal{L}(u_i^{\mathrm{p}}) + \rho\omega^2 u_i = 0 \tag{8.30}$$

Note that the second term of equation (8.30) contains the total displacement solution, u_i, in the domain which will be approximated using a set of global shape functions. We will come back to this later in this section. First, we write the boundary integral equation for the complementary function, u_i^c, in equation (8.29) using the static Green's function:

$$C_{ij}u_j^c - \int_\Gamma u_j^c t_{ij}^* \,d\Gamma + \int_\Gamma t_j^c u_{ij}^* \,d\Gamma = 0 \tag{8.31}$$

where u_{ij}^* and t_{ij}^* are the Kelvin's displacement and traction fundamental solutions, respectively, given by the equations (6.23) through (6.26). After discretization of the boundary Γ this boundary integral equation can be transformed into matrix BE equation:

$$[G]\{t^c) = [H]\{u^c\} \tag{8.32}$$

We can now substitute the total displacements and tractions for the complementary functions u^c and t^c appearing in this equation using equations (8.27) and (8.28), leading to:

$$[G]\{t\} - [H]\{u\} = [G]\{t^p) - [H]\{u^p\} \tag{8.33}$$

The particular solution u^p, appearing here, must be solved from the differential equation (8.30). To this end, total displacement solution u_i inside the domain Ω, which appears in the second term of equation (8.30), will be approximated, as it was done in the case of DRM, using an unknown fictitious density function $\{\Phi\}$ and a known global shape function f just as we did in equation (6.29):

$$u_i = \sum_{m=1}^{\infty} f_{ik}(x, \xi_m)\Phi_k(\xi_m) \tag{8.34}$$

This step is same as in equation (8.9) except for the global shape function notation, which is f here in place of C.

Given a specific form of the known functions f, equation (8.30) can be solved for the particular solution, u^p, and then the traction components, t^p, can be computed. It may be noted here that we are essentially approximating the inertia term of the differential equation in an attempt to compute the mass matrix. The computation of this term does not involve any spatial differentiation and as such it suffices most often to approximate this term with simpler functions compared to the stiffness terms. As done in the DRM, we will assume a set of global shape functions for these known functions f:

$$f_{ik}(x, \xi_m) = \delta_{ik}[R - r(x, \xi_m)] \tag{8.35}$$

where R can be taken as the largest distance between two points on the body and r is the distance between the field point x and the source point ξ_m. Upon substitution of the approximation (8.34) for the domain displacement, u_i, the particular integral differential equation (8.30) becomes:

$$\mathcal{L}\left(u_i^p\right) + \rho\omega^2 \sum_{m=1}^{\infty} f_{ik}(x, \xi_m)\Phi_k(\xi_m) = 0 \tag{8.36}$$

The particular solution, u^p, can now be chosen in such a way that it satisfies this equation. It can be written in the following manner:

$$u_i^p(x) = \sum_{m=1}^{\infty} D_{ik}(x, \xi_m)\Phi_k(\xi_m) \tag{8.37}$$

The displacement function, D, which satisfies equation (8.36) is found to be:

$$D_{ik}(x,\xi_m) = \frac{\rho\omega^2}{\mu}\left[(c_1 r - c_2 R)\,\delta_{ik} r^2 - c_3 y_i y_k r\right] \tag{8.38}$$

where:

$$y_i = x_i - \xi_{mi} \tag{8.39}$$

$$c_1 = \frac{2(d+3)(1-\nu)-1}{18(3d-1)(1-\nu)} \tag{8.40}$$

$$c_2 = \frac{1-2\nu}{2\{(1+d)-2\nu d\}} \tag{8.41}$$

$$c_3 = ½(1-\nu)(d^2+4d+3) \tag{8.42}$$

$d = 2$ for 2-D problems and 3 for 3-D problems. For 2-D problems, for example, D is given by:

$$D_{ik} = \frac{\rho\omega^2}{\mu}\left[\left\{\frac{(9-10\nu)}{90(1-\nu)}r - \frac{(1-2\nu)}{(6-8\nu)}R\right\}\delta_{ik}\,r^2 - \frac{1}{30(1-\nu)}y_i\,y_k r\right] \tag{8.43}$$

The traction components, t_i^{p}, corresponding to the particular integrals, u_i^{p}, of equation (8.37), can be computed using the strain-displacement and constitutive relations given in Chapter 5. Thus, the traction components, t_i^{p}, are found to be:

$$t_i^{\mathrm{p}}(x) = \sum_{m=1}^{\infty} T_{ik}(x,\xi_m)\Phi_k(\xi_m) \tag{8.44}$$

The traction functions are given by:

$$\begin{aligned} T_{ik}(x,\xi_m) &= \rho\omega^2\left[(c_4 r - c_5 R)\,y_k n_i + (c_6 r - 2c_2 R)\,y_i n_k\right] \\ &\quad + \left\{(c_6 r - 2c_2 R)\,\delta_{ik} - \frac{2c_3\,y_i\,y_k}{r}\right\} y_j n_j \end{aligned} \tag{8.45}$$

where:

$$c_4 = \frac{(d+3)\nu - 1}{3(3d-1)(1-\nu)} \tag{8.46}$$

$$c_5 = \frac{2\nu}{(1+d)-2\nu d} \tag{8.47}$$

$$c_6 = \frac{(d+2)-(d+3)\nu}{3(3d-1)(1-\nu)} \tag{8.48}$$

Thus, for 2-D problems, T would take the following form:

$$\begin{aligned} T_{ik} &= \rho\omega^2\left[\left\{\frac{(5\nu-1)}{15(1-\nu)}r - \frac{2\nu}{(3-4\nu)}R\right\} y_k n_i + \left\{\frac{(4-5\nu)}{15(1-\nu)}r - \frac{2(1-2\nu)}{2(3-4\nu)}R\right\} y_i n_k \right. \\ &\quad \left. + \left\{\left(\frac{(4-5\nu)}{15(1-\nu)}r - \frac{2(1-2\nu)}{2(3-4\nu)}R\right)\delta_{ik} - \frac{15(1-\nu)}{r}y_i y_k\right\} y_j n_j\right] \end{aligned} \tag{8.49}$$

We can obviously use the nodes on the boundary as the collocation points for the evaluation of the displacement and traction functions D and T respectively and write the equations (8.37) and (8.44) in the matrix form:

$$\{u^{\mathrm{p}}\} = \rho\omega^2[D]\{\Phi\} \tag{8.50}$$

$$\{t^{\mathrm{p}}\} = \rho\omega^2[T]\{\Phi\} \tag{8.51}$$

Substituting these discretized particular solutions of displacements and tractions back into equation (8.33) we arrive at the following:

$$[G]\{t\} - [H]\{u\} = \rho\omega^2([G][T] - [H][D])\{\Phi\} \tag{8.52}$$

Combining equations (8.34) and (8.35) we can write the total displacement approximation in terms of known functions and fictitious functions:

$$u_i(x) = \sum_{m=1}^{\infty} \delta_{ik}\,\{R - r(x, \xi_m)\}\,\Phi_k(\xi_m) \tag{8.53}$$

This relationship between the total displacement and fictitious function can be written in matrix form using the discretized boundary points as the collocation points:

$$\{u\} = [C]\{\Phi\} \tag{8.54}$$

We can solve for $\{\Phi\}$ from this relationship by inverting the matrix $[C]$:

$$\{\Phi\} = [C]^{-1}\{u\} \tag{8.55}$$

After substituting this value of $\{\Phi\}$ into equation (8.52), we can express the equilibrium equation entirely in terms of physical variables, such as, displacement and traction:

$$[G]\{t\} - [H]\{u\} = \rho\omega^2([G][T] - [H][D])[C]^{-1}\{u\} \tag{8.56}$$

This can be written as:

$$[G]\{t\} - [H]\{u\} = \rho\omega^2[M]\{u\} \tag{8.57}$$

where $[M]$ is the desired mass matrix:

$$[M] = ([G][T] - [H][D])[C]^{-1} \tag{8.58}$$

After applying appropriate displacement and traction boundary conditions, $u_i = 0$ and $t_i = 0$ respectively, we can cast equation (8.57) into a generalized eigenvalue problem:

$$[A]\{x_i\} = k^2[B]\{x_i\} \tag{8.59}$$

where:

$\{x_i\}$ are the eigenvectors $\lfloor u_i, t_i \rfloor$
k_i are the eigenvalues ($k_i = \omega_i/c$)

The natural frequencies of vibration of 2-D and 3-D elastic bodies can be computed by solving this generalized eigenvalue problem. Note that eigenvalue formulation for axisymmetric elastic bodies will require a separate formulation because a different fundamental solution will have to be used. The next section will present a number of 2-D and 3-D free vibration analysis examples in elasticity.

8.3.2. Example problems of eigenvalue analysis in elasticity

Unlike the finite element formulation, the system stiffness and mass matrices [A] and [B] respectively, resulting from boundary element formulation, are fully populated and unsymmetric. As a result, special eigenvalue extraction routines will be needed to compute frequencies and mode shapes from unsymmetric mass and stiffness matrices. Rajakumar and Rogers [178] and Rajakumar [179] developed the Lanczos algorithm, which is applied to unsymmetric generalized eigenvalue problems. This algorithm has been successfully applied to fully populated and unsymmetric mass and stiffness matrices generated by the BEM [131, 132, 136, 137]. Ahmad and Banerjee [123] used the eigenvalue extraction algorithm developed by Moller and Stewart [180] to solve for the frequencies of vibration of elastic bodies.

Example 8.5: Two-dimensional square and triangular elastic bodies

The triangular cantilevered planar elastic body is 10 units at the supported base with a span of eight units (Fig. 8.8). The material properties are taken as $E/\rho = 10{,}000$ and $\nu = 0.2$ (E = Young's Modulus, ρ = density and ν = Poisson's ratio). Different levels of boundary element discretizations are used to study the convergence behavior of the boundary element formulation. The results for first four modes are shown in Table 8.4. The time periods for the four modes from finite element runs with corresponding number of boundary nodes are found to be 0.430, 0.212, 0.192 and 0.125, respectively.

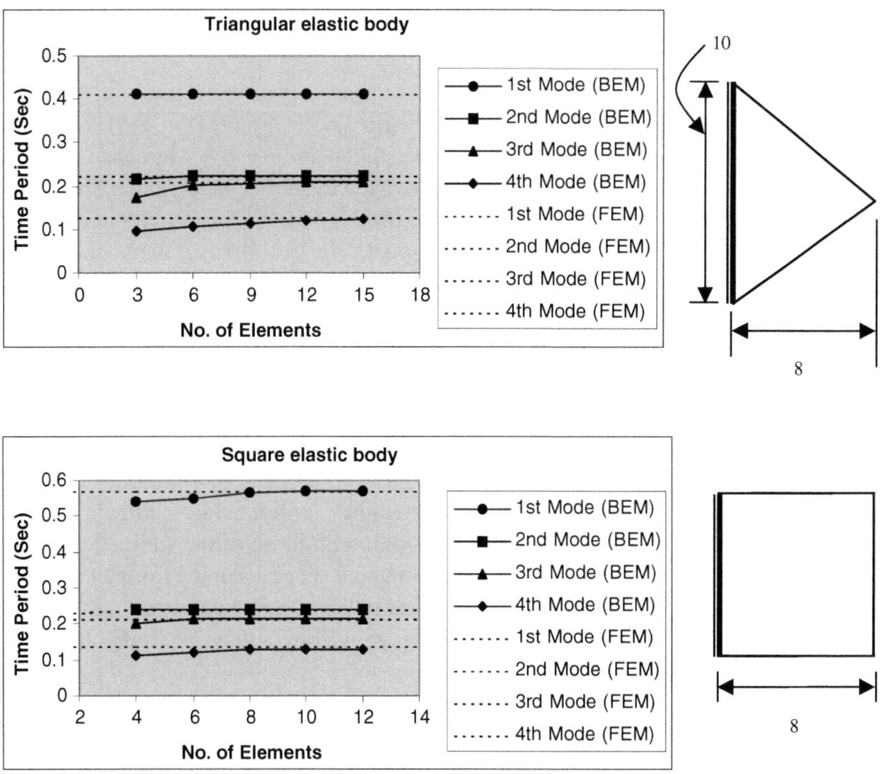

Figure 8.8. Convergence studies of time periods of planar square and triangular elastic bodies (FEM model with 121 nodes).

Table 8.4. Time periods of free vibration of triangular cantilevered elastic body.

Number of elements	Mode 1	Mode 2	Mode 3	Mode 4
3	0.432	0.207	0.138	0.081
6	0.430	0.212	0.180	0.095
9	0.430	0.212	0.189	0.104
12	0.430	0.212	0.191	0.109
15	0.430	0.212	0.192	0.111
18	0.430	0.212	0.192	0.112

Table 8.5. Time periods of free vibration of square cantilevered elastic body.

Number of elements	Mode 1	Mode 2	Mode 3	Mode 4
4	0.561	0.235	0.172	0.107
6	0.568	0.237	0.179	0.116
8	0.581	0.238	0.185	0.122
10	0.581	0.238	0.187	0.123
12	0.584	0.238	0.187	0.125
16	0.585	0.238	0.187	0.125

The square cantilevered planar elastic body is 6 units × 6 units (Fig. 8.8). The material properties are once again taken as $E/\rho = 10{,}000$ and $\nu = 0.2$. Here also different levels of boundary element discretizations are used to study convergence. The results for first four modes are shown in Table 8.5. The time periods from finite element runs are found to be 0.585, 0.238, 0.187 and 0.126, respectively. In both cases the boundary element solutions are seen to converge to the correct results for a small number of elements.

Example 8.6: Cantilever beam

The free vibration analysis is performed on a deep cantilevered beam of length = 24 units and height = 6 units (Fig. 8.9). The material properties are $E/\rho = 10{,}000$ and $\nu = 0.2$. A number of boundary element discretizations are used and the results are compared with finite element solutions. The first, second and fourth modes represent transverse vibration, whereas the third and fifth modes represent longitudinal vibration. The boundary element results compare very well with those of the finite element for the longitudinal modes. In order to obtain the same level of accuracy for the transverse modes, more boundary elements will be needed in the transverse direction. In general, the boundary element results are seen to compare well with those of finite elements.

Example 8.7: Shear wall

Next a shear wall with four openings is selected for study. The wall is modeled by BEM as well as FEM. The two discretizations are shown in Figure 8.10. The boundary element mesh consists of 29 quadratic line elements with 58 nodes, whereas the finite element mesh consists of 476 two-dimensional four-noded planar solid finite elements

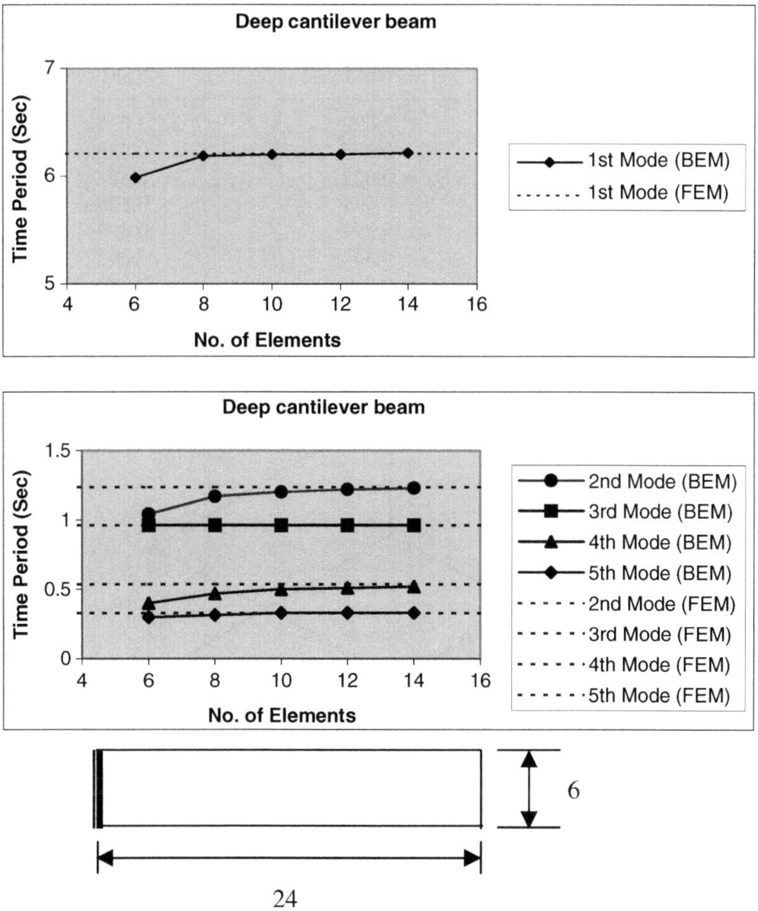

Figure 8.9. Convergence studies of time periods of vibrations of a deep cantilevered beam (FEM model with 451 nodes).

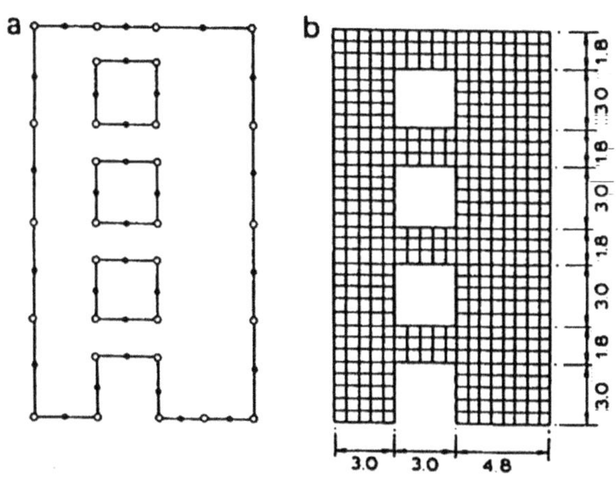

Figure 8.10. Two different discretizations of a shear wall: (a) boundary element model with 58 nodes and (b) finite element model with 559 nodes.

Table 8.6. Time periods of free vibrations for shear wall.

Mode	BEM	FEM
1	3.022	3.029
2	0.875	0.885
3	0.822	0.824
4	0.531	0.526
5	0.394	0.409
6	0.337	0.342
7	0.310	0.316
8	0.278	0.283

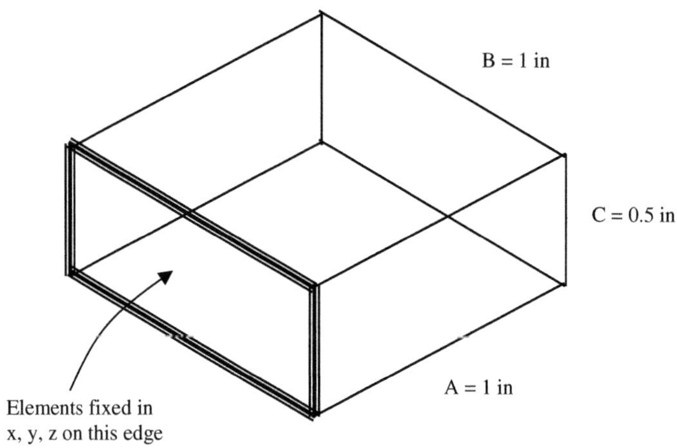

Figure 8.11. Natural frequency study of a rectangular parallelepiped.

with 559 nodes. The results for the boundary element and finite element analyses are shown in Table 8.6. It is remarkable that the boundary element results for this relatively complex geometry compare so closely with those of finite element even with a small number of nodes in BEM.

Example 8.8: Rectangular parallelepiped

A rectangular parallelepiped having dimensions $A = 1.0$, $B = 1.0$ and $C = 0.5$ are considered next for boundary element natural frequency analysis (Fig. 8.11). One of the edges is fixed in all directions and other edges are free. Eight-noded serendipity quadratic elements are utilized in the analysis [127]. Four different mesh densities, as shown in Table 8.7, are employed for the convergence study. The material properties of the parallelepiped are: Young's modulus $E = 16.126\text{E}6$ psi, Poisson's ratio $\nu = 0.3$ and density $\rho = 0.0007\,\text{lbm/in}^3$. The natural frequency ω is measured in radians/second. The results of the boundary element analyses are shown in Table 8.8. Four distinct modes of vibrations are manifested: EB = easy bending, SB = stiff bending, T = torsion and L = extension. The same problem was studied by Leissa and Zhang [181] where they used a Ritz technique to solve for the natural frequencies. Their solutions are known to be accurate within one per cent for the first mode.

Table 8.7. Different boundary element discretizations used for the rectangular parallelepiped.

Model	Elements along side … A	B	C	Number of elements	Number of nodes
1	1	1	1	6	20
2	1	2	1	10	32
3	1	3	1	14	44
4	2	3	1	22	68

Table 8.8. Convergence study of natural frequencies of a rectangular parallelepiped.

Mode	Type	Ritz	Mesh 1	(%)	Mesh 2	(%)	Mesh 3	(%)	Mesh 4	(%)
1	EB	0.447	0.472	5.5	0.429	−4.0	0.435	−2.7	0.442	−1.2
2	SB	0.667	0.664	−0.4	0.666	−0.3	0.668	0.1	0.661	−0.8
3	T	0.788	0.887	12.5	0.829	5.1	0.820	4.0	0.788	0.0
4	L	1.596	1.625	1.8	1.620	1.5	1.618	1.4	1.602	0.4
5	EB	1.664	2.136	28.3	1.797	8.0	1.729	3.9	1.689	1.5
6	SB	1.774			1.836	3.5	1.789	0.9	1.775	0.1
7	T	2.220			2.552	14.9	2.448	10.3	2.285	2.9
8	EB	2.278					3.033	33.3	2.365	3.8
9	L	2.797							2.842	1.6
10	SB	3.068							3.249	5.9

Mode type identification: EB = easy bending, SB = stiff bending, T = torsion and L = extension.

The boundary element analysis results in terms of the frequency parameter $\omega\sqrt{(\rho/E)}$ are listed in Table 8.8 where these are compared against those of the Ritz method. The first mode for each displacement type is seen to converge rapidly. More boundary elements will be needed in the longitudinal (side B) direction in order to achieve higher rate of convergence for the higher modes.

Example 8.9: Automobile crankshaft
In this example, a half symmetry model of an automobile crankshaft is considered [127]. It is modeled by eight-noded serendipity quadratic boundary elements as shown in Figure 8.12. The wider end was held by rigid lubricated rollers along outer edges and the rest of the surface was left free. The crankshaft was also modeled by finite elements in which the surface discretization was taken to be the same as the boundary element mesh. The results of the frequency analysis are presented in Table 8.9. The first, second and fourth mode frequencies are seen to be in good agreement.

Wilson, Miller and Banerjee [127] reports that the finite element solution failed to yield the third natural frequency for the given level of discretization. When the finite element mesh was refined, it yielded the third frequency as 47,600 Hz, leaving the first, second and fourth frequency essentially unchanged.

It can be pointed out here that there is a subtle but important distinction between the free vibration problem in elasticity and that in acoustics. The mode shapes in the elasticity free vibration problem are somewhat controlled by the problem boundary,

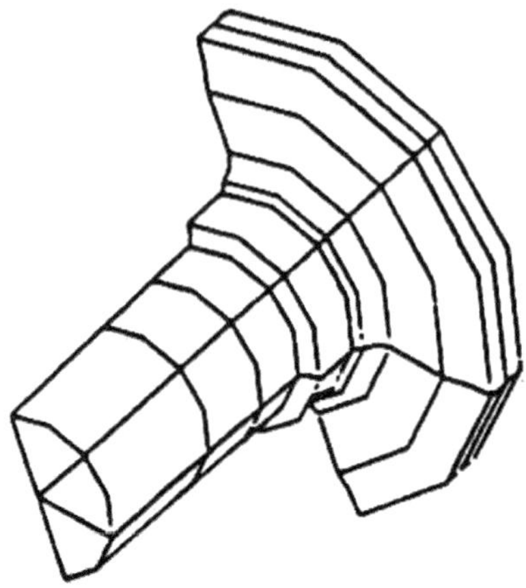

Figure 8.12. Boundary element mesh of automobile crankshaft.

Table 8.9. Natural frequencies (Hz) of automobile crankshaft.

Mode	Finite element solution	Boundary element solution
1	17,400	18,200
2	35,800	34,300
3	–	47,500
4	66,300	66,300

the domain being in tune with the boundary because of compatibility conditions. On the contrary, in the case of acoustics, the problem boundary does influence eigenmodes, but it does not control them to the same extent as in the elasticity problem. Rather eigenmodes are governed here by the continuity of the eigenfunctions. As a result, compared to elasticity problems, more internal collocation points will be required to accurately solve truly 2-D or 3-D acoustic eigenvalue problems having complex eigenfunctions. The method of introducing internal collocation points in the domain will be developed in Chapter 9.

Chapter 9

Advanced Concepts in Boundary Element Algebraic Eigenproblem

9.1. Introduction

The previous two chapters laid the foundation for the formulation of boundary element algebraic eigenvalue problem. In Chapter 7, we showed the basic idea of setting up the algebraic eigenproblem in boundary elements employing the Internal Cell Method (ICM). In Chapter 8, Dual Reciprocity Method (DRM) and Particular Integral Method (PIM) were developed by extending the ICM. Unlike in ICM, the power of DRM and PIM lies in the fact that the discretization remains confined to the boundary only without the need for breaking up the domain into internal cells.

Let us summarize here the ideas behind these methods: DRM separates the free vibration differential equation into two components such that one component is free of frequency parameter and the other term contains the frequency parameter. The former leads to the stiffness-type matrix, whereas the latter, which is the inertia term, leads to system mass matrix. The component without the frequency parameter is transformed into boundary integral equation using a fundamental solution, which is free of the frequency, the so-called static Green's function. The other component, which is the inertia term, becomes a volume integral just like a body force term in the boundary element method (BEM). The dependent variable in the volume integral, pressure in acoustics or displacement in elasticity, is expressed in terms of a global shape function (GSF) and a fictitious density function. The volume integral is then transformed into boundary integrals with the help of Gauss's divergence theorem. The free vibration problem is thus cast into a generalized eigensystem with boundary only discretization.

In PIM, the set up of the algebraic eigenvalue problem is approached along the same lines as DRM with a slight difference. The same global shape functions are used here to approximate the inertia term. However, rather than using Gauss's divergence theorem, the inertia term is forced to satisfy the original free vibration differential equation in an effort to extract a particular integral. Thus, PIM uses the concept of solving differential equations by means of a complementary function and a particular integral. The matrix manipulations involved in PIM are similar to those in DRM and lead to identical algebraic eigenproblem equations (8.22) and (8.59).

Both DRM and PIM can be used to solve free vibration problems in elasticity and acoustics. A few refinements of DRM and PIM will be shown in this chapter. The method presented here offers certain advantages over DRM and PIM. This method may be designated as "Fictitious Function Method" (FFM). It is primarily based on DRM and PIM. The idea behind this method is to temporarily retain the fictitious density function $\{\Phi\}$ in equations (8.16) and (8.52) as the system variable instead of the original dependent variable, pressure $(P\}$ in acoustics or displacement $\{u\}$ in elasticity.

Note that the relationship between the original dependent variable and the fictitious function [eqns. (8.17) and (8.54)] represents a linear transformation. The eigenvalues of the system remain unchanged under this change of variables. The eigenvectors in terms of the original variable can be retrieved very easily through equations (8.17) and (8.54). As a result of the use of the fictitious density function rather than the original dependent variable as the system variable, we can avoid the matrix inversion shown in equations (8.18) and (8.55).

FFM is particularly helpful for the commonly encountered pure Neumann problem ($Q=\partial P/\partial n=0$) in acoustics. However, it is also applicable to problem where a part of the boundary is of the Dirichlet type ($P=0$ or $u=0$). In this case, a small size matrix needs to be inverted. The same observation would remain true for the fluid–structure coupled problems in which the boundary conditions on the wall are given by $Q=\partial P/\partial n=\rho\omega^2 U_n$. The eigenvalue formulation details for the fluid–structure coupled problem will be presented in Chapter 10.

9.2. Algebraic eigenvalue formulation using fictitious function method

This method utilizes the property of matrices that the eigenvalues of a matrix remain the same under a linear transformation of its eigenvectors. The eigenfrequencies are extracted along with the transformed eigenvectors. Because the inversion of a matrix of size as large as the system matrix is avoided, FFM is computationally more efficient for large problems with Neumann boundary conditions. Details of FFM are presented below, followed by an example problem.

We develop the FFM here starting with the PIM presented in Chapter 8. The problem of acoustic eigenanalysis was posed in equations (8.1) through (8.4). In the PIM we saw that the total pressure is split into two components, a complementary function pressure and a particular solution pressure:

$$P = P^{\mathrm{C}} + P^{\mathrm{I}} \quad \text{and} \quad Q = \partial P/\partial n = \partial P^{\mathrm{C}}/\partial n + \partial P^{\mathrm{I}}/\partial n \tag{9.1}$$

The complementary function pressure satisfies the Laplacian portion of the governing equation:

$$\nabla^2 P^{\mathrm{C}} = 0 \tag{9.2}$$

and the particular solution pressure will satisfy

$$\nabla^2 P^{\mathrm{I}} = -k^2 P \tag{9.3}$$

The boundary integral statement of equation (9.2) is set up using the static Green's function (frequency-independent fundamental solution) approach as given in equations (2.8) and (2.9). Here we rewrite the resulting Boundary Element matrix equation for complimentary function pressure:

$$[G]\{Q^{\mathrm{C}}\} = [H]\{P^{\mathrm{C}}\} \tag{9.4}$$

The complementary function pressure is eliminated from this equation using the relationships given in equations (9.1):

$$[G]\{Q\} - [H]\{P\} = [G]\{Q^{\mathrm{I}}\} - [H]\{P^{\mathrm{I}}\} \tag{9.5}$$

The particular integrals P^{I} must be obtained by solving equation (9.3). To this end, the pressure P inside the domain is approximated using a GSF and fictitious density functions, shown in equations (8.9) and (8.10). Substituting for the total pressure from equation (8.9) into equation (9.3), we obtain:

$$\nabla^2 P^{\mathrm{I}} + k^2 \sum_{m=1}^{\infty} (R - r)\, \Phi(\xi^m) = 0 \tag{9.6}$$

The particular integral pressure, P^{I}, can be solved from this equation employing one of the differential equation solution methods, such as, either a trial and error method or the method of undetermined coefficients. The following particular integral solution satisfies the above equation and, therefore, is a solution to this equation:

$$P^{\mathrm{I}}(x) = k^2 \sum_{m=1}^{\infty} D(x, \xi^m)\Phi(\xi^m) = 0 \tag{9.7}$$

The normal gradient $Q^{\mathrm{I}} = \partial P^{\mathrm{I}}/\partial n$ for the particular integral P^{I} can now be computed. Let us write the particular integral solutions in matrix form:

$$\{P^{\mathrm{I}}\} = k^2[D]\{\Phi\} \quad \text{and} \quad \{Q^{\mathrm{I}}\} = k^2[T]\{\Phi\} \tag{9.8}$$

Substituting these into equation (9.5), we obtain the algebraic eigenvalue problem in terms of both the pressure P and the fictitious density function Φ:

$$[G]\{Q\} - [H]\{P\} = k^2([G][T] - [H][D])\{\Phi\} \tag{9.9}$$

This is rewritten as:

$$[G]\{Q\} - [H]\{P\} = k^2[\overline{M}]\{\Phi\} \tag{9.10}$$

with this definition of mass-type matrix:

$$[\overline{M}] = ([G][T] - [H][D]) \tag{9.11}$$

In standard DRM or PIM, the fictitious density functions $\{\Phi\}$ are eliminated in favor of the physical pressures $\{P\}$ using the relation given in equation (8.17). See Chapter 8 for further details on standard DRM and PIM.

For acoustic eigenvalue analysis in particular, most frequently the task is to compute resonant frequencies of closed cavities with acoustically rigid boundaries. In this case, it can be seen from equations (9.10) and (8.17) that, if the eigenvalue problem is posed in terms of $\{\Phi\}$ rather than $\{P\}$, the inversion of the $[C]$ matrix can be avoided, leaving the eigenvalues unchanged. Here we work with a set of fictitious density functions $\{\Phi\}$ only temporarily, as the eigenmodes in terms of $\{P\}$ can be easily retrieved using equation (8.17). Detailed derivations of FFM for different boundary conditions are given below.

9.2.1. Fictitious function method with pure Neumann boundary condition

Consider the case in which all the boundary walls of the enclosure are acoustically hard, i.e., $Q = \partial P/\partial n = 0$ for all nodes on the boundary. Then equation (9.9) becomes:

$$-[H]\{P\} = k^2([G][T] - [H][D])\{\Phi\} \tag{9.12}$$

Substituting the values of $\{P\}$ from equation (8.17) into the above, the entire formulation can be posed in terms of the fictitious density functions $\{\Phi\}$:

$$-[H][C]\{\Phi\} = k^2([G][T] - [H][D])\{\Phi\} \tag{9.13}$$

Let us rewrite this equation in compact form:

$$[\overline{K}]\{\Phi\} = k^2[\overline{M}]\{\Phi\} \tag{9.14}$$

with the following definition of stiffness-type matrix:

$$[\overline{K}] = -[H][C] \tag{9.15}$$

The eigenvalues $\lambda = k^2$ of this eigensystem is the same as that of equation (8.19). The only difference is that equation (9.14) does not require any matrix inversion. The eigenvectors in terms of physical variable $\{P\}$ can be obtained from equation (8.17), if desired.

9.2.2. Fictitious function method with mixed boundary conditions

In case the boundary conditions on the walls of the enclosures are mixed with both Neumann and Dirichlet type conditions:

$$P_1 = 0 \quad \text{on } \Gamma_1 \qquad \text{and} \qquad (\partial P/\partial n)_2 = Q_2 = 0 \quad \text{on } \Gamma_2 \tag{9.16}$$

equations (9.10) and (8.17) can be partitioned as:

$$\begin{bmatrix} G_{11} & G_{12} \\ G_{21} & G_{22} \end{bmatrix} \begin{Bmatrix} Q_1 \\ Q_2 \end{Bmatrix} - \begin{bmatrix} H_{11} & H_{12} \\ H_{21} & H_{22} \end{bmatrix} \begin{Bmatrix} P_1 \\ P_2 \end{Bmatrix} = k^2 \begin{bmatrix} \overline{M}_{11} & \overline{M}_{12} \\ \overline{M}_{21} & \overline{M}_{22} \end{bmatrix} \begin{Bmatrix} \Phi_1 \\ \Phi_2 \end{Bmatrix} \tag{9.17}$$

$$\begin{Bmatrix} P_1 \\ P_2 \end{Bmatrix} = \begin{bmatrix} C_{11} & C_{12} \\ C_{21} & C_{22} \end{bmatrix} \begin{Bmatrix} \Phi_1 \\ \Phi_2 \end{Bmatrix} \tag{9.18}$$

$n_1 + n_2 = n$, total number of nodes on the boundary. The above partitioned equations can be turned into:

$$\begin{bmatrix} G_{11} & H_{12}\left(C_{21}C_{11}^{-1}C_{12} - C_{22}\right) \\ G_{21} & H_{22}\left(C_{21}C_{11}^{-1}C_{12} - C_{22}\right) \end{bmatrix} \begin{Bmatrix} Q_1 \\ \Phi_2 \end{Bmatrix} = k^2 \begin{bmatrix} 0 & \left(\overline{M}_{12} - \overline{M}_{11}C_{11}^{-1}C_{12}\right) \\ 0 & \left(\overline{M}_{22} - \overline{M}_{21}C_{11}^{-1}C_{12}\right) \end{bmatrix} \begin{Bmatrix} Q_1 \\ \Phi_2 \end{Bmatrix} \tag{9.19}$$

$$\{P_2\} = \left(C_{22} - C_{21}C_{11}^{-1}C_{12}\right)\{\Phi_2\} \tag{9.20}$$

Equations (9.19) and (9.20) can be used to extract eigenvalues and eigenvectors for mixed boundary conditions. We note here that FFM would not be efficient for large values of n_2.

9.2.3. Fictitious function method in fluid–structure interaction problem

Fictitious Function Method can also be applied to the eigenvalue analysis of coupled fluid–structure problems. The Neumann boundary condition takes the form given in equation (8.4), $\partial P/\partial n = \rho\omega^2 u_n$, when the acoustic fluid is in contact with a vibrating structure. Using this fluid–structure boundary condition in equation (9.10) one can write:

$$\omega^2[\overline{G}]\{u\} - [H]\{P\} = k^2[\overline{M}]\{\Phi\} \tag{9.21}$$

This can be rewritten in the following form:

$$\begin{bmatrix} 0 & 0 \\ 0 & -[H][C] \end{bmatrix} \begin{Bmatrix} u \\ \Phi \end{Bmatrix} = k^2 \begin{bmatrix} 0 & 0 \\ -c^2[\overline{G}] & [\overline{M}] \end{bmatrix} \begin{Bmatrix} u \\ \Phi \end{Bmatrix} \tag{9.22}$$

where:

$$[\overline{G}] = \rho[Gn_x|Gn_y] \tag{9.23}$$

n_x and n_y respectively, are the x and y components of the unit normal at the interface boundary and $\{u\}$ are the structural displacements. The matrix equation for the vibrating structure can be written down, accounting for the fluid pressure load at the fluid–structure interface, as follows:

$$-\omega^2[M_s]\{u\} + [K_s]\{u\} = -[R]\{P\} \tag{9.24}$$

With the help of equation (8.17), equation (9.24) can be written as:

$$\begin{bmatrix} [K_s] & -[R][C] \\ 0 & 0 \end{bmatrix} \begin{Bmatrix} u \\ \Phi \end{Bmatrix} = k^2 \begin{bmatrix} c^2[M_s] & 0 \\ 0 & 0 \end{bmatrix} \begin{Bmatrix} u \\ \Phi \end{Bmatrix} \tag{9.25}$$

where $[K_s]$ and $[M_s]$ are the structural stiffness and mass matrices, respectively. $[R]$ is the coupling matrix representing the surface area on which the fluid pressure load acts on the structure at the fluid–structure interface. Equations (9.22) and (9.25) can be combined together in the form:

$$\begin{bmatrix} [K_s] & -[R][C] \\ 0 & -[H][C] \end{bmatrix} \begin{Bmatrix} u \\ \Phi \end{Bmatrix} = k^2 \begin{bmatrix} c^2[M_s] & 0 \\ -c^2[\overline{G}] & [\overline{M}] \end{bmatrix} \begin{Bmatrix} u \\ \Phi \end{Bmatrix} \tag{9.26}$$

This equation can be used to study an interesting class of eigenvalue problems involving acoustic fluid–structure interaction. This subject will be further studied in details in Chapter 10 where example problems for fluid–structure interaction will be discussed.

9.3. Example problems using fictitious function method

Example 9.1: Impedance tube

The impedance tube problem was considered in Chapter 8 in the context of standard DRM. Here the same problem is used to illustrate FFM. Acoustics laboratories commonly use this set-up for experimental purposes (Fig. 9.1). The boundary walls of the impedance tube are assumed to be acoustically hard with $\partial P/\partial n = 0$. The speed of sound "$c$" in the air is taken as 340 m/s. The tube length and width are chosen to be $a = 40$ m and $b = 2$ m, respectively. Sixteen linear elements are used along the length and two elements along the width of the tube.

The results of the analysis are shown in Table 9.1 where the theoretical solutions are also shown for comparison. The results from FFM are seen to be in good agreement with the theoretical solutions. The eigenmodes are presented in Figure 9.2 in terms of physical pressures as well as fictitious function Φ. For this one-dimensional problem,

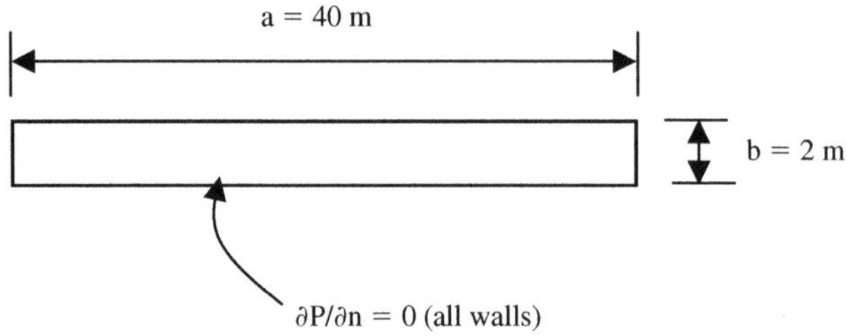

Figure 9.1. Impedance tube (speed of sound $c = 340$ m/s).

Table 9.1. Comparison of eigenvalues for impedance tube.

Mode	FFM (Hz)	Theory (Hz)
1	0	0
2	4.26	4.25
3	8.53	8.50
4	12.91	12.75
5	17.30	17.0

the variation of Φ is seen to closely follow that of P. The fifth mode containing two full sine waves is marginally captured here with only four linear segments per half sine wave.

9.4. Effect of internal collocation points on eigensolutions

It was pointed out earlier that unlike in BE elastic eigenvalue analysis, the BE acoustic eigenvalue analysis will require internal collocation points for accurate solutions. The global shape functions, given by equation (8.10), used to approximate the total pressure in DRM as well as PIM are not adequate in representing true two-dimensional (2-D) or three-dimensional (3-D) eigenmodes. In order to prove the need for internal collocation points for eigenfrequency computation, let us consider the extreme case of a circular acoustic enclosure (Fig. 9.3) with Neumann boundary conditions, $\partial P/\partial n = 0$, on the boundary wall. Theoretical eigenfrequency solutions for this circular domain Neumann problem are given by the characteristic equation:

$$J'_m(ka) = 0 \quad (m = 0, 1, 2, \ldots) \tag{9.27}$$

$J'_m(\,)$ is the first derivative of the m-th order Bessel's function of the first kind. The circular boundary is discretized first using 36 linear segments. The boundary is next discretized using quadratic elements with the same number of nodal points. The results are presented in Table 9.2, where the theoretical solutions as well as solutions from ANSYS [160] acoustic finite element run are also presented. It can be seen that the quadratic representation of the boundary improves the solution slightly, but they are still poor. It means that no matter how good the boundary discretization, the boundary element acoustic eigenvalue analysis without internal points will yield very poor results in this case, even for the fundamental mode. Furthermore, it fails to yield the ring mode solution. It happens because all the modes for the circular domain are truly 2-D, which

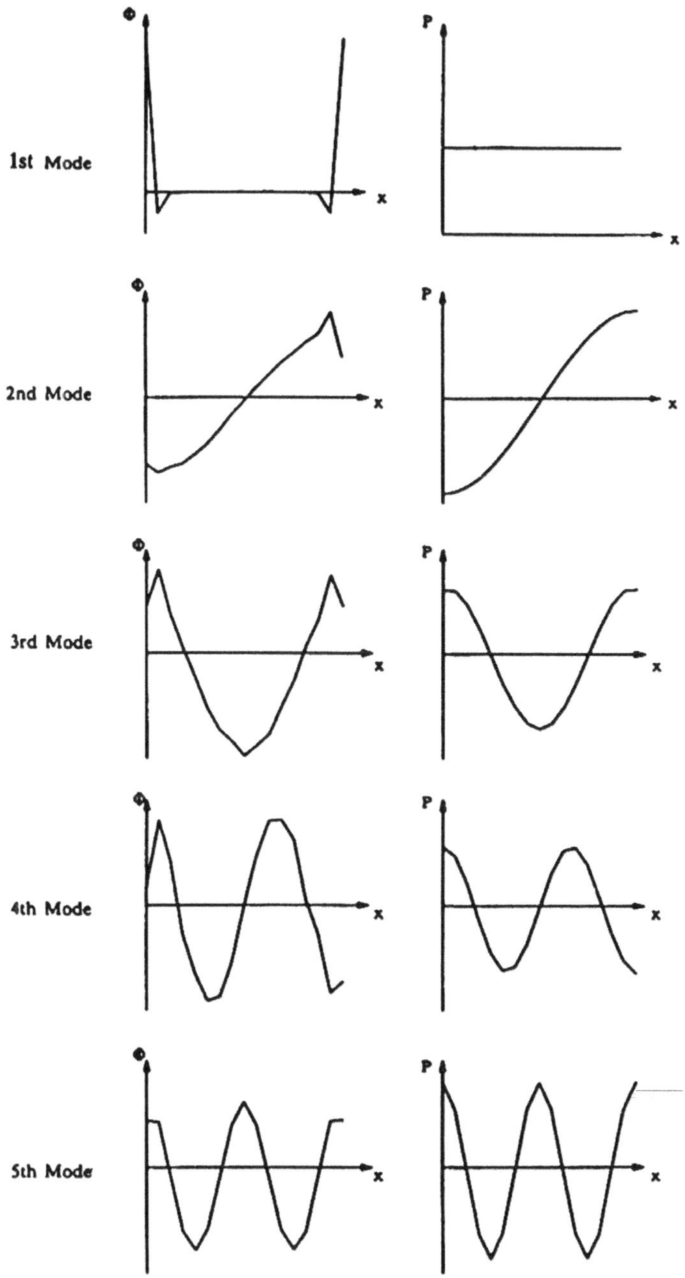

Figure 9.2. Eigenmodes of the impedance tube.

means that the shape of eigenfunctions can have dramatic variations within the domain and boundary nodes alone are not capable of describing the modes adequately. This observation is obviously true for DRM, PIM and FFM.

Here we will show how to insert internal collocation points in the domain [116, 131] and how to include these internal domain pressure variables as part of the total

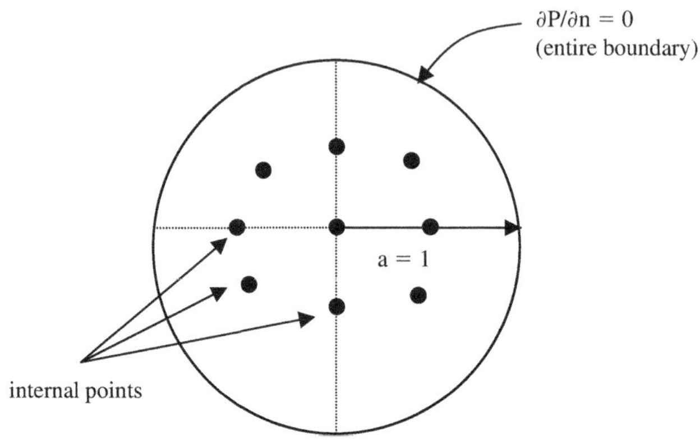

Figure 9.3. Circular acoustic enclosure (speed of sound $c = 340$ m/s).

Table 9.2. Eigenvalues of circular domain without internal points.

Mode	DRM/PIM/FFM Linear elements (Hz)	DRM/PIM/FFM Quadratic elements (Hz)	Finite element (216 elements) (Hz)	Theory (Hz)
1	105.15	104.89	100.02	99.63
2	105.15	104.89	100.02	99.63
3	177.10	176.12	166.33	165.27
4	177.10	176.12	166.33	165.27
5	–	–	209.40	207.34
6	247.44	246.09	229.88	227.34
7	247.44	246.09	229.88	227.34

problem. Note that the eigensolutions can also be improved by dividing the domain into several regions. See Chapter 4 for zoning procedure in BEM.

In Chapter 2 we showed that after the solution of potential and its normal derivatives on the boundary collocation points, the solution inside the domain can be found using equations (2.28) and (2.29). In the present case, the pressure $\{\hat{P}\}$ inside the domain can be computed using the pressure $\{P\}$ and pressure gradients $\{q\}$ on the boundary:

$$[I]\{\hat{P}\} + [\hat{H}]\{P\} = [\hat{G}]\{Q\} \tag{9.28}$$

$[I]$ is an identity matrix of order which equals to the number of internal points used. These internal point solutions can be combined with the boundary solutions given in equation (9.13) to yield:

$$-[\tilde{H}][\tilde{C}]\{\tilde{\Phi}\} = k^2([\tilde{G}][\tilde{T}] - [\tilde{H}][\tilde{D}])\{\tilde{\Phi}\} \tag{9.29}$$

where:

$$[\tilde{H}] = \begin{bmatrix} [H] & [0] \\ [\hat{H}] & [I] \end{bmatrix}; \quad [\tilde{G}] = \begin{bmatrix} [G] \\ [\hat{G}] \end{bmatrix}; \quad \{\tilde{\Phi}\} = \begin{Bmatrix} \Phi \\ \hat{\Phi} \end{Bmatrix} \tag{9.30}$$

$\{\hat{\Phi}\}$ represents the additional fictitious functions for the internal nodes corresponding to the internal pressures $\{\hat{P}\}$. $[\tilde{C}]$, $[\tilde{D}]$ and $[\tilde{T}]$ are the augmented matrices containing the

Table 9.3. Eigenvalues of circular domain with and without internal points (linear elements).

Mode	No internal points (Hz)	One internal point (Hz)	Five internal points (Hz)	Nine internal points (Hz)
1	105.15	105.15	102.42	101.50
2	105.15	105.15	102.42	101.50
3	177.10	177.10	172.23	172.23
4	177.10	177.10	177.10	172.23
5	–	239.12	225.76	221.97
6	247.44	247.44	246.05	243.66
7	247.44	247.44	246.05	243.66

Table 9.4. Eigenvalues of circular domain with and without internal points (quadratic elements).

Mode	No internal points (Hz)	One internal point (Hz)	Five internal points (Hz)	Nine internal points (Hz)
1	104.89	104.89	102.19	101.27
2	104.89	104.89	102.19	101.27
3	176.12	176.12	171.45	171.45
4	176.12	176.12	176.12	171.45
5	–	239.22	225.76	221.97
6	246.09	246.09	244.75	242.42
7	246.09	246.09	244.75	242.42

coefficients of internal points as well as the boundary nodes corresponding to $[C]$, $[D]$ and $[T]$ matrices, respectively. Equation (9.29) will yield more accurate eigensolutions because of the representation of eigenfunctions on the boundary as well as inside the domain. The next section will present example problems to demonstrate the effects of inclusion of internal collocation points in the problem.

9.4.1. Examples to show effect of internal collocation points on eigensolutions

Example 9.2: Circular acoustic domain with internal points

Let us consider the same circular domain problem. This time it is solved using varying number of internal collocation points. The results are presented in Table 9.3 for linear elements and Table 9.4 for quadratic elements. It can be seen that the results improve considerably with the gradual addition of internal node points. The ring mode solution appears with the addition of only one internal point at the center of the domain.

Example 9.3: Trapezoidal acoustic domain

A trapezoidal acoustic domain is considered next (Fig. 9.4). Finite element solutions as well as experimental results for this problem are reported by Shuku and Ishihara [176]. This problem is solved using FFM without and with internal points. The results are presented in Table 9.5 where the finite element and experimental solutions are also included for comparison. In this case, the first mode is essentially one-dimensional and the solution appears to be reasonable for this mode without the use of internal

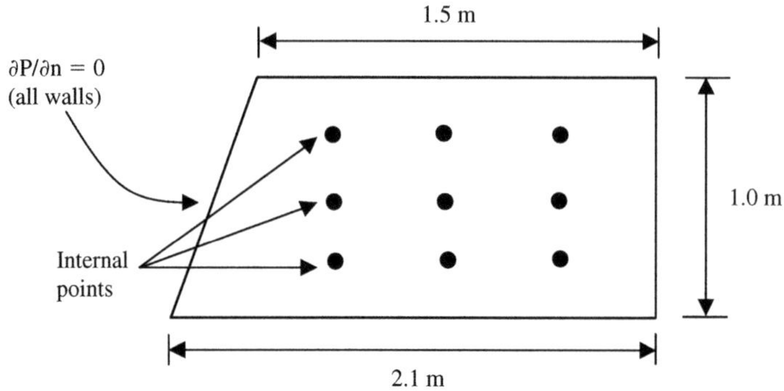

Figure 9.4. Trapezoidal acoustic enclosure (speed of sound $c = 340$ m/s).

Table 9.5. Eigenvalues for trapezoidal model.

	Eigenvalues (Hz)			
Mode	No internal points	Nine internal points	Experimental	Finite element
1	93.7	92.9	93	92.5
2	169.3	165.0	164	162.5
3	187.6	182.3	182	179.1

node points. But the GSF represented by the boundary nodes alone is found to be inadequate for the second and third modes. The inclusion of nine internal nodes improved the accuracy considerably.

It is apparent that computer codes written to perform eigenvalue analysis of acoustic enclosures must always keep provision to include internal points or to break the domain into a number of zones in order to obtain reliable eigensolutions.

9.5. Polynomial-based particular integral method

The DRMs and PIMs of Chapter 8, and their variants presented in this chapter, entail the important step in which the dependent variable, such as pressure, contained in the inertia term is approximated using a global shape function. We chose the global shape functions given in equations (8.10) and (8.35) in Chapter 8 with only brief discussion about alternative shape functions that could also be considered for this purpose. Here we will investigate some of the other applicable GSFs. The selection of GSF is guided by several important considerations:

(a) First of all, in the context of DRM, the chosen function must allow the integral transformation from domain to boundary via the application of Gauss's divergence theorem to equation (8.6).
(b) Secondly, in the context of PIM, the chosen function must be such that we can find suitable particular solutions for equation (8.30).

(c) And finally, the functions must be chosen such that they produce sufficiently accurate eigensolution, and at the same time, can be generated in a computationally efficient manner.

The inertia term approximation with GSFs used in equations (8.10) and (8.35) can be improved by the addition of constant and/or higher order terms to the shape functions [116]. The shape function $C(x, \xi_m)$, given in equation (8.10), can be generalized further to contain terms with positive powers of the distance between the field and source points, $r(x, \xi_m)$, rather than just the linear term $[R - r(x, \xi_m)]$ alone. Here, we consider a complete polynomial as a possible choice for the GSF. So, the acoustic pressure that appears in the inertia term can be represented by [135]:

$$P(x) = \sum_{m=1}^{\infty} C_m(x)\Phi_m \tag{9.31}$$

where $C_m(x)$ is a set of polynomial functions and Φ_m is the vector of unknown coefficients associated with these polynomial functions. However, the above equation is an infinite series. On practical grounds, we replace it by a finite series containing a finite number of polynomial terms. For example, a complete second order polynomial expression in 2-D and 3-D spaces can be written down as finite series equations shown below:

$$P(x) = \Phi_1 + x_1\Phi_2 + x_2\Phi_3 + x_1^2\Phi_4 + x_2^2\Phi_5 + x_1x_2\Phi_6 \qquad \text{(2-D)} \tag{9.32}$$

$$\begin{aligned} P(x) &= \Phi_1 + x_1\Phi_2 + x_2\Phi_3 + x_3\Phi_4 + x_1^2\Phi_5 + x_2^2\Phi_6 + x_3^2\Phi_7 \\ &\quad + x_1x_2\Phi_8 + x_2x_3\Phi_9 + x_1x_3\Phi_{10} \qquad \text{(3-D)} \end{aligned} \tag{9.33}$$

Here, the total number of unknown coefficients is six for 2-D and 10 for 3-D spatial representations of the pressure in the domain. When the linear GSF $[R - r(x, \xi_m)]$ is used to approximate the dependent variable in the inertia term using the boundary nodes as collocation points, the number of unknown coefficients (Φ_m) equals the number of boundary nodes, N. This leads to a square matrix $[C]$, in equation (8.17), which is inverted [eqn. (8.18)] to solve for the unknown coefficients vector $\{\Phi\}$. Here, when polynomials in equation (9.32) or (9.33) are used to approximate the dependent variable, the number of unknown coefficients remains six for 2-D and 10 for 3-D. This usually will lead to a rectangular $[C]$ matrix since, typically, there will be more collocation points (i.e. boundary nodes) than the unknown coefficients where the dependent variable is collocated.

Therefore, a least square regression approach needs to be employed to compute the unknown coefficients in this over-determined system. The least square method may be employed in the following way:

(a) Representing the collocation of pressure, we can write the relations given in equations (9.32) and (9.33) in matrix form as $[P] = [C]\{\Phi\}$. This is same as equation (8.17), except here $[C]$ is a rectangular matrix.
(b) Pre-multiplying this equation by $[C]^T$ yields $[C]^T[P] = [C]^T[C]\{\Phi\}$. The resulting matrix $[C]^T[C]$ is square and, therefore, can be inverted. The product $[C]^T[C]$ yields a 6×6 matrix for 2-D and 10×10 for 3-D for the second-order polynomial functions considered here. For the linear GSF based approximation, the $[C]$ matrix in equation (8.17) is of size $N \times N$, where N is the number of boundary nodes.

(c) Now, the coefficient vector $\{\Phi\}$ is computed as:

$$\{\Phi\} = ([C]^T[C])^{-1}[C]^T\{P\} = [\overline{C}]\{P\} \tag{9.34}$$

Notice here that the size of the matrix to be inverted is much smaller (either 6×6 or 10×10) compared to the case of linear GSFs.

We can now proceed to formulate the algebraic eigenvalue problem using PIM as presented in Section 8.3.1. To this end, we substitute the polynomial-based approximation from equation (9.31) into equation (9.3) to obtain:

$$\nabla^2 P^{\mathrm{I}} + k^2 \sum_{m=1}^{\infty} C_m(x)\Phi_m = 0 \tag{9.35}$$

where $C_m(x)$ are the polynomial terms, $(1, x_1, x_2, x_1^2, \dots)$, found in equations (9.32) for 2-D and (9.33) for 3-D. We can now solve for the particular solution $P^{\mathrm{I}}(x)$:

$$P^{\mathrm{I}}(x) = k^2 \sum_{m=1}^{M} D_m(x)\Phi_m \tag{9.36}$$

$M = 6$ for 2-D and 10 for 3-D for the second-order polynomial functions used here. $D_m(x)$ for 3-D are given below:

$$\left.\begin{aligned}
D_1 &= -\frac{1}{2d}\sum_{i=1}^{3} x_i x_i \\
D_m &= -\frac{1}{2(d+2)}\sum_{i=1}^{3} x_i x_i x_j && (m = 2, 3, 4; j = m-1) \\
D_m &= -\frac{1}{12}x_j^4 && (m = 5, 6, 7; j = m-4) \\
D_m &= -\frac{1}{2(d+4)}\sum_{i=1}^{3} x_i x_i x_j x_k && (m = 8; j = 1, k = 2 \\
& && m = 9; j = 1, k = 3 \\
& && m = 10; j = 2, k = 3)
\end{aligned}\right\} \tag{9.37}$$

The normal derivative of this particular solution can be found as:

$$Q^{\mathrm{I}} = \frac{\partial P^{\mathrm{I}}(x)}{\partial n} = k^2 \sum_{m=1}^{10} T_m(x)\Phi_m \tag{9.38}$$

$T_m(x)$ are given below:

$$\left.\begin{aligned}
T_1 &= -\frac{1}{d}\sum_{i=1}^{3} x_i n_i \\
T_m &= -\frac{1}{2(d+2)}\left\{\sum_{i=1}^{3}(2x_i n_i x_j + x_i x_i n_j)\right\} && (m = 2, 3, 4; j = m-1) \\
T_m &= -\frac{1}{3}x_j^3 n_j && (m = 5, 6, 7; j = m-4) \\
T_m &= -\frac{1}{2(d+4)}\left[\sum_{i=1}^{3}\{2x_i n_i x_j x_k + x_i x_i(x_j n_k + x_k n_j)\}\right] \\
&\quad (m = 8; j = 1, k = 2 \\
&\quad m = 9; j = 1, k = 3 \\
&\quad m = 10; j = 2, k = 3)
\end{aligned}\right\} \tag{9.39}$$

In the above particular solution expressions, $d = 3$ for 3-D ($d = 2$ for 2-D), n_i are the components of outward normal at the boundary nodes. The solution for the 2-D case can be deduced from above by putting $i = 1, 2$ and setting the third-direction components to zero. After boundary discretization, the above particular solutions can be written in matrix form as in equation (9.8) with the difference that $[D]$ and $[T]$ are now rectangular matrices. Substitution of these matrix equations for particular solution and its derivative into equation (9.5) leads to equation (9.9). We can then eliminate the unknown coefficients $\{\Phi\}$ from equation (9.9) using (9.34) and arrive at the following eigenvalue problem:

$$[G]\{Q\} - [H]\{P\} = k^2([G][T] - [H][D])[\overline{C}]\{P\} \tag{9.40}$$

As before, after applying the boundary conditions, equation (9.40) can be cast into a generalized algebraic eigenvalue problem, $[A]\{x_i\} = k^2[B]\{x_i\}$, where $\{x_i\}$ are the eigenvectors $\lfloor P_i, Q_i \rfloor$, k_i are the eigenvalues ($k_i = \omega_i / c$) and $Q_i = \partial P_i / \partial n$.

Next, we will illustrate the use of this polynomial-based PIM with a number of example problems. In all the examples the geometry, pressure, and particular solution variables are approximated using isoparametric quadratic shape functions. The boundary walls for all the presented problems are considered as acoustically hard surfaces ($Q = \partial P / \partial n = 0$) and the speed of sound is taken as 340 m/s. The eigenvalue problem presented by the above equation was solved using EISPACK [182] employing an extraction technique based on the Arnoldi's algorithm [127, 135].

9.5.1. Examples of polynomial-based particular integral method

Example 9.4: Two-dimensional rectangular acoustic cavity

This impedance tube problem has been presented earlier (see example problems 8.1 and 9.1). Here eigensolutions from polynomial shape function (PSF) approximation are compared against those from GSF approximation. The results are shown in Table 9.6 where the analytical solutions are also noted. Since this is essentially a one-dimensional problem, the break-up of the domain into a number of regions does not have any perceptible influence on the solution.

Example 9.5: Automotive compartment with and without seats

This problem was presented in Chapter 8. See example problem 8.3 and Figures 8.3 through 8.6 and Table 8.2. Here Table 9.7 compares the GSF approximation solutions with those from PSF approximation. Finite element and experimental results are also shown in the table. Since the method based on PSF approximation requires inversion of a much smaller matrix, it requires significantly less CPU time compared to the GSF approximation based technique.

Table 9.6. Acoustic resonant frequencies (Hz) for impedance tube.

		GSF			PSF	
Mode	Analytical	One region	Two regions	Four regions	Two regions	Four regions
1	4.25	4.24	4.24	4.22	4.25	4.25
2	8.5	8.46	8.48	8.45	8.73	8.51
3	12.75	12.68	12.69	12.70	13.32	12.80
4	17.0	16.88	16.89	16.97	18.17	17.73

Table 9.7. Acoustic resonant frequencies (Hz) for a hatchback car interior compartment.

	Without seats					With seats				
			GSF		PSF			GSF		PSF
Mode	Experimental	FEM	One region	Four regions	four regions	Experimental	FEM	One region	Four regions	Four regions
1	60	68	69	72	73	53	50	48	49	49
2	110	105	104	110	111	–	79	75	78	78
3	135	152	153	155	155	–	125	116	122	129
4	–	179	190	186	193	–	163	157	159	166

Table 9.8. Acoustic resonant frequencies (Hz) of rectilinear cavity.

		GSF		PSF	GSF	PSF
Mode	Analytical	One region	Four regions	Four regions	Four regions	Four regions
1	10.625	10.607	10.591	10.634	10.505	10.634
2	21.25	21.120	21.162	21.834	21.057	21.291
3	31.875	31.590	31.619	33.840	31.659	32.098
4	42.5	41.947	41.938	46.440	42.241	44.524

Table 9.9. Acoustic resonant frequencies (Hz) of a truck-cab.

Mode	FEM	GSF One region	PSF One region	GSF Three regions	PSF Three regions
1	67	72	78	72	72
2	82	85	92	85	91
3	105	109	124	111	119

Example 9.6: Three-dimensional rectilinear acoustic cavity
This is a 3-D version of the impedance tube problem and has the closed form analytical solution given by equation (8.23), that is, although the geometry is 3-D, it is still essentially a one-dimensional problem. Let us consider a 16 meter long tube with a one meter square cross-section. The problem is solved using polynomial based shape function approximation as well as GSF approximation. The results along with the analytical solutions are presented in Table 9.8. The tube is treated as a single-region and four-region BE domain. As noted earlier, since this is a 1-D problem, the use of multiple regions does not act in favor of improving the eigensolution.

Example 9.7: Three-dimensional truck-cab acoustic model
See example 8.4 for the geometric description of this 3-D acoustic eigenvalue problem. Here two different BE models are used to solve the eigenproblem. The first case is a single-region BE model as described in example 8.4. In the second case, the 3-D cavity is broken into three BE zones: each of the end two zones is discretized using 30 boundary elements and 90 nodes whereas the middle zone is modeled using 25 elements and 75 nodes. Table 9.9 lists the results, where the results from GSF based approximation as well as finite element method (FEM) [177] are also tabulated.

9.6. Multiple reciprocity method (MRM)

The MRM can be looked upon as an extension of the idea of DRM presented in Section 8.2. The purpose of DRM and MRM is to transform domain integrals arising from the application of BEM to free-vibration problems, problems with body forces, and treatment of certain classes of non-linearities. The DRM requires that the dependent variable appearing inside the inertia term be approximated by functions having certain qualities so that Gauss's divergence theorem may be applied to transform the domain integral to boundary integrals. Here, MRM eliminates the need to approximate the dependent variable. Instead, in this method, Gauss's divergence theorem is repeatedly applied to the domain integral term using higher order Green's functions until the domain term becomes negligible [138–140]. Let us consider the Poisson's equation for the purpose of illustrating the method:

$$\nabla^2 P + b_0(x) = 0 \tag{9.41}$$

where $b_0(x)$ is a function representing body force. If P_0^* represents the fundamental solution to the Lapalace's equation, $\nabla^2 P = 0$, then we can write the following integral equation:

$$C(\xi)P(\xi) + \int_\Gamma P(x)Q_0^*(x,\xi)\mathrm{d}\Gamma(x) - \int_\Gamma P_0^*(x,\xi)Q(x)\mathrm{d}\Gamma(x) = \int_\Omega P_0^*(x,\xi)b_0(x)\mathrm{d}\Omega \tag{9.42}$$

The concern here is to transform the domain integral body force term into boundary integral. To this end, let us consider the following sequence of body force functions and fundamental solutions:

$$b_{j+1}(x) = \nabla^2 b_j(x) \quad (j = 0, 1, 2, 3, \ldots) \tag{9.43}$$

$$\nabla^2 P_{j+1}^* = P_j^*(x) \quad (j = 0, 1, 2, 3, \ldots) \tag{9.44}$$

These relationships make the repeated application of Gauss's divergence theorem to the domain integral possible. The series of particular solutions of equation (9.44) are known as the higher-order fundamental solutions. Making use of these series of functions, the domain integral term of equation (9.42) can be evaluated as:

$$\int_\Omega P_0^*(x,\xi)b_0(x)\,\mathrm{d}\Omega = \sum_{j=0}^{N}\int_\Gamma \frac{\partial P_{j+1}^*(x,\xi)}{\partial n}b_0(x)\,\mathrm{d}\Gamma - \sum_{j=0}^{N}\int_\Gamma \frac{\partial b_{j+1}^*(x)}{\partial n}P_{j+1}(x,\xi)\,\mathrm{d}\Gamma$$
$$+ \int_\Omega P_{n+1}^*(x,\xi)b_{n+1}(x)\,\mathrm{d}\Omega \tag{9.45}$$

The assumption here is that if we apply the Gauss's divergence theorem a sufficient number of times, then the domain integral term of equation (9.45) will become negligible. For example, if the body force term $b_0(x)$ is a polynomial of $x, b_1(x), b_2(x), \ldots$, then the repeated application of divergence theorem will diminish their order. In that case, we can evaluate the domain integral with sufficient accuracy by a finite number of terms. The higher-order fundamental solution series $P_j^*(x)$ for equation (9.44) are given by:

$$P_j{}^*(x,\xi) = -\frac{1}{2\pi}\left[r(x,\xi)\right]^{2j}\frac{1}{4^j\,(j!)^2}\left[\ln\{r(x,\xi)\} - \sum_{m=1}^{j}\frac{1}{m}\right] \quad \text{(2-D)} \tag{9.46}$$

$$P_j{}^*(x,\xi) = \frac{1}{4\pi r(x,\xi)}\frac{1}{(2j)!}\left[r(x,\xi)\right]^{2j} \quad \text{(3-D)} \tag{9.47}$$

In the case of the Helmholtz equation, we have $b_0(x) = k^2P(x)$ so that $b_1(x) = \nabla^2 b_0(x) = k^2\nabla^2P(x) = (-k^2)^2P(x)$, and so on, which leads to:

$$b_j(x) = (-k^2)^{j+1}P(x) \tag{9.48}$$

We can substitute this into equation (9.45) to evaluate the domain integral term for the Helmholtz equation and as a result we can rewrite integral equation (9.42) for the Helmholtz equation in the following manner:

$$C(\xi)P(\xi) + \sum_{j=0}^{N}(-k^2)^j\int_\Gamma P(x)Q_j{}^*(x,\xi)\,\mathrm{d}\Gamma(x) - \sum_{j=0}^{N}(-k^2)^j\int_\Gamma Q(x)P_j{}^*(x,\xi)\,\mathrm{d}\Gamma(x)$$
$$= (-1)^N(k^2)^{N+1}\int_\Omega P(x)P_N{}^*(x,\xi)\,\mathrm{d}\Omega \tag{9.49}$$

It can be seen from the fundamental solutions (9.46) and (9.47) that the leftover domain integral term in equation (9.49) will tend to zero for sufficiently large N. We, therefore, obtain the boundary-only integral equation representation for the Helmholtz equation:

$$C(\xi)P(\xi) + \sum_{j=0}^{N}(-k^2)^j \int_{\Gamma} P(x)Q_j{}^*(x,\xi)\,\mathrm{d}\Gamma(x)$$

$$-\sum_{j=0}^{N}(-k^2)^j \int_{\Gamma} Q(x)P_j{}^*(x,\xi)\,\mathrm{d}\Gamma(x) = 0 \tag{9.50}$$

Discretizing the boundary Γ and performing integration on the boundary element segments, the integral equation can be converted to:

$$\left[\sum_{j=0}^{N}(-k^2)^j[H_j]\right]\{P\} = \left[\sum_{j=0}^{N}(-k^2)^j[G_j]\right]\{Q\} \tag{9.51}$$

where $Q = \partial P/\partial n$. After applying specified homogeneous boundary conditions into equation (9.51) and assembling the matrix, one obtains:

$$[A(k)]\{x\} = [B(k)]\{0\} \tag{9.52}$$

where:

$$[A(k)] = [A_0] + k^2[A_1] + k^4[A_2] + \cdots + k^{2N}[A_N] \tag{9.53}$$

Kamiya and Andoh [141] solved equation (9.52) using Newton–Raphson iteration along with LU decomposition [183]. Note that the matrices $[A_0] \cdots [A_N]$ do not contain the frequency parameter k and, therefore, need not be formed at each iteration. Additionally, with this technique, internal collocation points are not required for an accurate evaluation of the resonant frequencies. Since it is an iterative technique, the solution procedure requires an initial rough estimate of the frequencies at the start of the iteration process. The Newton–Raphson solution technique may be looked upon as an enhanced determinant search method (DSM).

Example 9.8: An illustration of MRM: a rectangular parallelopiped acoustic cavity problem

As an example of the application of MRM with Newton–Raphson/LU iteration procedure, consider a rectangular parallelopiped [143], shown in Figure 9.5. The dimensions, boundary conditions and boundary discretizations are shown in the figure. Two different discretizations, shown in Figures 9.5(a) and 9.5(b), respectively, are considered: one with 24 constant elements and the other with 94 constant elements. The analytical solution for this problem is given by:

$$k = \pi\sqrt{\left(\frac{m}{L_x}\right)^2 + \left(\frac{n}{L_y}\right)^2 + \left(\frac{t+\frac{1}{2}}{L_z}\right)^2} \quad (m, n, t = 0, 1, 2, \ldots) \tag{9.54}$$

The results of the analysis are presented in Table 9.10 where the closed-form solutions are also shown. The maximum error is seen to diminish from five per cent to two per cent as the discretization gets finer.

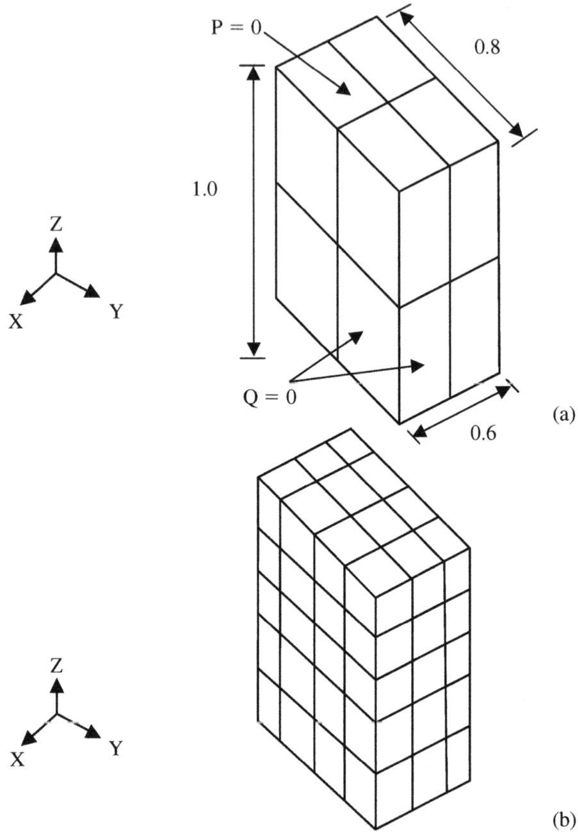

Figure 9.5. A rectangular parallelopiped acoustical cavity. (a) 24 elements; (b) 94 elements.

Table 9.10. Results for rectangular parallelopiped acoustical cavity.

(m, n, t)	Analytical solution	MRM 24 elements	Error (%)	MRM 94 elements	Error (%)
(0, 0, 0)	1.57	1.64	4.5	1.59	1.3
(0, 1, 0)	4.23	4.44	5.0	4.28	1.2
(0, 0, 1)	4.71	4.76	1.1	4.71	0.0
(1, 0, 0)	5.47	5.60	2.4	5.58	2.0

9.6.1. MRM and matrix augmentation

As noted above, MRM does not lead to an algebraic eigenvalue formulation, but it rather falls in the category of enhanced DSM. However, we can use a matrix augmentation procedure [142] to cast equation (9.52) into a generalized eigenproblem. To facilitate it, let us rewrite equation (9.52) in the following manner with the help of equation (9.53):

$$[A_0]\{x_0\} + [A_1]\{x_1\} + [A_2]\{x_2\} + \cdots + [A_N]\{x_N\} = \{0\} \tag{9.55}$$

where the right-hand side of equation (9.52) is basically a null vector and it is set equal to zero in equation (9.55). Let us also define $k^2 = \lambda$ and introduce a series of vector relationships as follows:

$$\{x_i\} = \lambda^i\{x_0\} \quad (i = 0, 1, 2, \ldots, N) \tag{9.56}$$

If we put $i = 0$ in the above relation, we get the following:

$$\{x_0\} = \{x_0\} \tag{9.57}$$

The equations (9.56) and (9.57) can be turned into a recursive relation of the form:

$$\{x_{i+1}\} = \lambda\{x_i\} \quad (i = 0, 1, 2, \ldots, N-1) \tag{9.58}$$

Equations (9.55) and (9.58) can be combined into a single augmented matrix equation:

$$[\overline{A}]\{\overline{x}\} = \lambda[\hat{A}]\{\overline{x}\} \tag{9.59}$$

The augmented matrices $[\overline{A}]$ and $[\hat{A}]$, and the augmented vector $\{\overline{x}\}$ are then given by:

$$[\overline{A}] = \begin{bmatrix} [A_{N-1}] & [A_{N-2}] & \cdots & \cdots & [A_1] & [A_0] \\ [I] & [0] & \cdots & \cdots & [0] & [0] \\ & & \ddots & & & \\ & & & \ddots & & \\ & & & & \ddots & \\ & & & [I] & [0] & [0] \\ & & & & [I] & [0] \end{bmatrix} \tag{9.60}$$

$$[\hat{A}] = \begin{bmatrix} -[A_N] & [0] & \cdots & \cdots & [0] & [0] \\ [0] & [I] & \cdots & \cdots & [0] & [0] \\ & & \ddots & & & \\ & & & \ddots & & \\ & & & & \ddots & \\ & & & [0] & [I] & [0] \\ & & & & [0] & [I] \end{bmatrix} \tag{9.61}$$

$$\{\overline{x}\} = \{\{x_{N-1}\}\{x_{N-2}\}\cdots\{x_0\}\}^T \tag{9.62}$$

Now the problem could be solved using generalized eigensolvers. As can be seen from the matrices (9.60) and (9.61), the order of the matrices to be solved has gone up significantly, requiring more computer time and storage. While it was motivated by the apparent advantage of being able to solve the eigenproblem using a generalized eigensolver as a black-box, the matrix augmentation procedure may defeat the very purpose of using boundary-only discretization technique in the first place. The only remaining advantage of BE eigenvalue formulation using this approach will then be in the preparation of meshes for complex geometries since putting internal points or breaking up into regions is not needed for MRM.

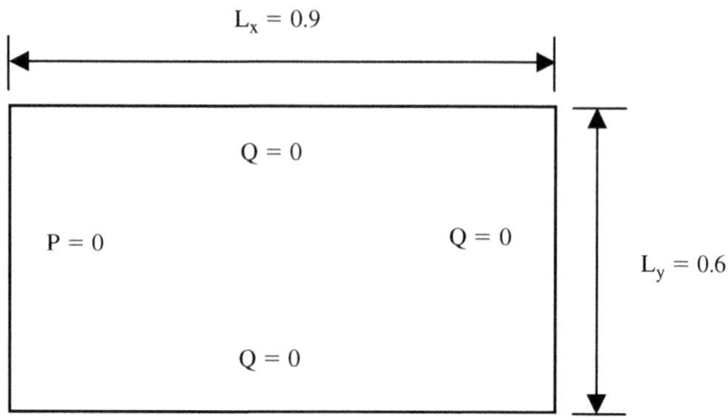

Figure 9.6. A rectangular acoustical cavity.

Table 9.11. Results for rectangular acoustical cavity.

MRM with matrix augmentation		MRM with LU/Newton		
26 Elements	44 Elements	26 Elements	44 Elements	Analytical solution
1.75	1.75	1.74	1.74	1.75
4.81*	4.81*			
5.25	5.24	5.25	5.25	5.24

*Fictitious eigenvalue.

Let us solve a 2-D problem for illustration. The problem is a 0.9×0.6 rectangular acoustical cavity with mixed boundary conditions, as shown in Figure 9.6. Two discretizations, one with 26 constant elements and the other with 44 constant elements, were used to solve the problem. The closed-form analytical solution for this problem is given by:

$$k = \pi\sqrt{\left(\frac{m}{L_x}\right)^2 + \left(\frac{n}{L_y}\right)^2} \quad (m, n = 0, 1, 2, \ldots) \tag{9.63}$$

The results are presented in Table 9.11 where the analytical solutions are compared against those obtained using (a) MRM with Newton/LU iterations and (b) MRM with matrix augmentation [143]. Both results agree very well with the closed-form solutions. The second row in Table 9.11, shown with an asterisk, is a fictitious eigenvalue, which is not predicted by the closed-form solution. According to Kamiya et al. [143], the fictitious eigenvalues correspond to the zeros of the determinant of the matrix $[B(k)]$ in equation (9.52). In practical applications, again, a robust method of detecting these spurious modes will have to be developed for this method to be useful.

Chen and co-workers (see [155]) observed that the MRM formulation, presented above, deals only with the real part of the complete complex-valued MRM formulation. Since the formulation leaves out the imaginary part, it results into insufficient number of constraints for the eigenequation, introducing spurious roots into the eigenproblem. Chen and his co-workers used residual method and singular value decomposition technique [155] to filter out these spurious modes from the eigensolution.

9.7. Series expansion methods (SEM) with matrix augmentation

It is clear that the use of a fundamental solution which rigorously corresponds to a governing differential equation leads to an eigenvalue formulation with the frequency parameter implicitly embedded into the system matrices. See for example the characteristic equations (6.51) and (6.55). This forced us to use direct DSM as described in Section 6.4.1. In Section 6.4.2, we also presented an enhanced DSM in which we expanded the characteristic equation into a series. The frequency parameters are factored out of the system matrices by this expansion [see eqns. (6.60) and (6.61)]. We can now use a matrix augmentation process, as outlined in the last section, to recast this series equation into a generalized eigenproblem, such as equation (9.59).

It can be shown that the series equation, viz., equation (6.60), is equivalent to equation (9.52) derived using MRM [143, 184, 185]. Note that the SEM [144] uses a fundamental solution for the Helmholtz equation for acoustic problems, whereas the MRM uses higher-order fundamental solutions for the Laplacian portion of Helmholtz equation for the same problem.

Chapter 10

Acoustic Fluid–Structure Interaction Problems

10.1. Introduction

In this chapter the eigenvalue analysis of acoustic fluid–structure systems encountered in acoustical cavities with flexible structure boundaries, such as a fluid-filled container or an automobile cabin enclosure, is considered. In applications involving acoustic cavities with flexible wall boundaries, the computation of the structural and cavity resonance involves solving the acoustic fluid–structure eigenproblem. Typically, the finite element method (FEM) is used to solve such fluid–structure coupled eigenproblems [178, 186]. However, when the problem size gets larger, as in the case of a structure in contact with a large extent of fluid, the finite element (FE) discretized stiffness and mass matrices of the coupled problem become very large, significantly increasing the eigenvalue computation time. In these situations, the boundary element method (BEM) becomes attractive since the discretization of the acoustic fluid domain leads to placing fluid nodes and elements only on the wetted surface of the structure, thus leading to relatively smaller size matrices for the coupled problem. However, boundary element (BE) matrices are non-symmetric; so also, are the pressure-displacement based finite element matrices even though the system sub-matrices are symmetric. Efficient methods of computing the eigenvalues and eigenvectors of non-symmetric eigenvalue problems will be presented in Chapter 11.

We will formulate the fluid–structure eigenproblem using boundary element discretization of the fluid and finite element discretization of the structure. For the fluid domain, any of the acoustic fluid boundary element eigenanalysis formulations such as Dual Reciprocity (DRM), Multiple Reciprocity (MRM) or Particular Integral Method, (PIM) presented in Chapters 8 and 9 can be used. For the structure, finite element eigenformulation is employed. In Section 9.2.3, the boundary element–finite element fluid–structure formulation was briefly outlined. However, we presented only the acoustic cavity eigenanalysis results there. In this chapter, eigenanalysis of the fluid–structure problem is presented in detail. We also present a number of illustrative problems where we compute the eigenfrequencies of structures in contact with fluid.

Boundary element–finite element coupled fluid–structure analysis is quite commonly employed in the frequency domain analysis of fluid–structure problems, where the coupled system response to excitation forces is computed as outlined in Section 5.2.1. Setting the excitation forces to zero, it is also possible to solve for the natural frequencies of the coupled problem using the method of determinant search to compute the eigenfrequencies and modes as shown in Sections 6.2 and 6.4. In this chapter, however, the boundary element–finite element coupled problem is set up as an algebraic eigenvalue problem in order to extract the eigenfrequencies and mode shapes of the fluid–structure system. To our knowledge, the authors of this book made the first ever attempt to perform a boundary element–finite element coupled algebraic eigenvalue analysis for acoustic fluid–structure problem [137].

10.2. Boundary element–finite element coupled eigenanalysis of fluid–structure system

We consider a fluid–structure system shown schematically in Figure 10.1. The fluid domain is confined and has two different types of boundaries. One portion of the fluid boundary is acoustically hard and the other portion is in contact with a flexible structure. For the time harmonic acoustic pressure, $p = Pe^{j\omega t}$, in the fluid, the wave equation reduces to the Helmholtz equation $\nabla^2 P + k^2 P = 0$, where P is the pressure amplitude, $k = \omega/c$ is the wave number, c is the speed of sound in the fluid and ω is the circular frequency of pressure oscillations. The boundary conditions of the fluid boundary in terms of the pressure amplitude can be written as:
Acoustically hard boundary:

$$\partial P/\partial n = 0 \tag{10.1a}$$

Fluid–Structure Interface:

$$\partial P/\partial n = \rho\omega^2 \boldsymbol{u} \cdot \boldsymbol{n} \tag{10.1b}$$

where $\boldsymbol{u}$ is the displacement vector of the structure at the interface, $\boldsymbol{n}$ is the unit normal of the interface drawn into the structure, and ρ is the fluid density and "·" represents the dot product.

The boundary element algebraic eigenvalue problem of the Helmholtz equation for the fluid can be formulated using either the DRM or the PIM presented in Chapter 8. Both the methods yield the same end result. The reader is referred to Chapter 8 for detailed derivations of the algebraic eigenvalue problem using these methods. The

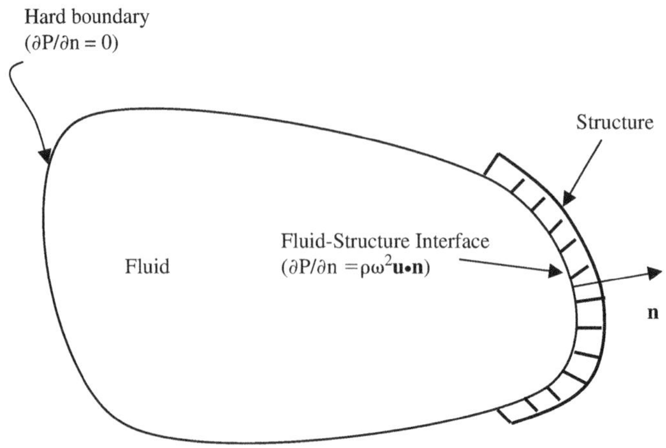

Figure 10.1. Acoustic fluid–structure problem.

application of any of these methods leads to the following equation [eqn. (8.16) in DRM, eqn. (8.52) in PIM)]:

$$[G]\{\partial P/\partial n\} - [H]\{P\} = k^2([G][T] - [H][D])\{\Phi\} \tag{10.2}$$

The matrices appearing in this equation are defined in Chapter 8. Here, it is possible to replace the fictitious function vector $\{\Phi\}$ in favor of the physical variable $\{P\}$ using equation (8.17). However, here we choose to retain $\{\Phi\}$ as the unknown vector in order to avoid the inversion of the $[C]$ matrix, as shown in the Fictitious Function Method (FFM) in Section 9.2. Once $\{\Phi\}$ is known, which will be part of the eigenvector solution, $\{P\}$ can be recovered using equation (8.17). For acoustic cavities with hard boundaries all around, the algebraic eigenproblem is obtained by setting $\{\partial P/\partial n\} = 0$ in equation (10.2). When the boundary is not hard, i.e., when the boundary is flexible, the fluid–structure interface condition $\{\partial P/\partial n\} = \rho\omega^2 \boldsymbol{u} \cdot \boldsymbol{n}$ is used to couple the acoustic domain to the oscillating structure.

If we denote the x and y components of the unit normal, $\boldsymbol{n}$, as n_x and n_y, we can write the boundary integral matrix equation (10.2) in terms of the interface nodal displacements $\{u\}$ and nodal fictitious functions $\{\Phi\}$ as unknowns.

$$\rho\omega^2[Gn_x|Gn_y]\{u\} - [H][C]\{\Phi\} = (\omega/c)^2([G][T] - [H][D])\{\Phi\} \tag{10.3}$$

The finite element structural dynamic equation can now be written taking into account the pressure force of the acoustic fluid acting on the structure [186].

$$[K_s]\{u\} - \omega^2[M_s]\{u\} = [R]\{P\} \tag{10.4}$$

The matrix $[R]$ represents the surface area at the interface, and $[K_s]$ and $[M_s]$ are the structural stiffness and mass matrices, respectively. Substituting for $\{P\}$ from equation (8.17), the structural dynamic equation (10.4) is rewritten as:

$$[K_s]\{u\} - [R][C]\{\Phi\} = \omega^2[M_s]\{u\} \tag{10.5}$$

The boundary element–finite element coupled eigenproblem can be written combining equations (10.3) and (10.5):

$$\begin{bmatrix} K_s & -[R][C] \\ 0 & -[H][C] \end{bmatrix} \begin{Bmatrix} u \\ \Phi \end{Bmatrix} = \omega^2 \begin{bmatrix} [M_s] & 0 \\ -\rho[Gn_x|Gn_y] & \frac{1}{c^2}([G][T] - [H][D]) \end{bmatrix} \begin{Bmatrix} u \\ \Phi \end{Bmatrix} \tag{10.6}$$

The matrices of the eigenproblem given in equation (10.6) are non-symmetric. Furthermore, the acoustic fluid boundary element matrices in the bottom row are fully populated. The effectiveness of the fluid–structure eigenanalysis will, therefore, depend upon an efficient method for extracting the eigenvalues of this non-symmetric system. In Chapter 11, we will present eigenvalue computational procedures based on the Lanczos method for non-symmetric matrices. The Lanczos subspace method is a powerful computational tool for extracting a few eigenvalues of a large non-symmetric eigenproblem within a specified range of its eigenvalue spectrum. Usually, for the fluid–structure problem the lowest resonant frequencies are of interest, and the Lanczos method is well suited for that.

As pointed out earlier, the pressure-displacement based finite element–finite element coupled eigenproblem is also non-symmetric, and will involve much larger sized fluid matrices than the boundary element–finite element eigenproblem for the same given fluid–structure analysis. Therefore, eigenproblem formulation using the

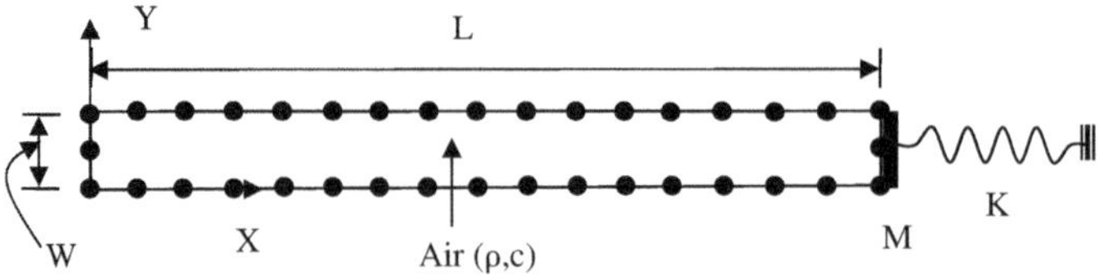

Figure 10.2. Impedance tube coupled to spring-mass oscillator ($L = 40$ m, $W = 2$ m, $K = 60$ N/m, $M = 0.25$ kg, $\rho = 1.12$ kg/m^3, $c = 340$ m/s).

boundary element–finite element coupled analysis approach, given by equation (10.6), would be relatively more efficient.

10.2.1. Application of fluid–structure eigenformulation

The example problems in this section are intended to show the application of fluid–structure eigenanalysis, and to compare the effectiveness of the boundary element–finite element eigenanalysis with respect to the finite element–finite element coupled eigenanalysis [178, 186]. First, through a simple example, we validate the boundary element–finite element coupled formulation by comparing the computed results with the theoretical solution. Then, practical applications of the method are illustrated by considering two representative fluid–structure problems.

Example 10.1: Impedance tube coupled to spring-mass oscillator

This one-dimensional (1-D) problem is a spring-mass oscillator placed at one end of an acoustic tube as shown in Figure 10.2. The other end of the air-filled tube is closed. Tube dimensions are chosen with a large length to width ratio, L/W, so that the coupled system behaves, essentially, as a 1-D acoustic tube oscillator in the lower range of frequencies that we are interested. The mass, M, and stiffness, K, of the oscillator are chosen to be light and soft compared to the mass and stiffness, respectively, of the air column in the tube.

For such an oscillator–tube combination the resonant frequencies are given by an approximate expression [187] of the form:

$$\omega_n = \frac{(2n-1)\pi c}{2L} \tag{10.7}$$

This approximate expression is derived from the exact frequency expression:

$$\cot(\omega_n L/c) - a\omega_n L/c + bc/(\omega_n L) = 0 \tag{10.8}$$

The coefficients a and b are the mass and stiffness ratios given by $M/(\rho AL)$ and $KL/(\rho Ac^2)$, respectively, where A is the cross-section area of the tube and L is its length. ρ and c are the density of the fluid and the speed of sound in the fluid, respectively. The theoretical frequencies listed in Table 10.1 have been evaluated using the exact frequency expression given by equation (10.8).

For the boundary element–finite element coupled eigenanalysis, the fluid domain boundary is discretized into 36 linear boundary elements as shown in Figure 10.2. The mass M which acts as the piston is modeled using stiff beam finite elements, and the spring K is modeled by a discrete stiffness element. The eigenvalues of the unsymmetric eigenvalue problem were extracted using the Lanczos eigenvalue solver, presented in Chapter 11, and the eigenfrequencies are listed in Table 10.1. The computed frequencies agree very well with the theoretical frequencies. Also, presented in Table 10.1 are

Table 10.1. Resonant frequencies of the impedance tube coupled to oscillator (Hz).

Mode	Theoretical	BE–FE	FE–FE
1	2.127	2.1167124	2.1199338
2	6.360	6.3803167	6.3801376
3	10.597	10.674617	10.701530
4	14.835	15.054761	15.125397
5	19.073	19.536871	19.693362
6	23.211	24.205379	24.446556
7	27.549	29.067250	29.423477
8	31.787	34.244301	34.655527
9	36.025	39.689099	40.158698
10	40.263	45.558443	45.919184

the eigenfrequencies computed using the finite element–finite element coupled analysis using 16 bilinear two-dimensional (2-D) acoustic fluid elements, available in the ANSYS program [160], along the length of the tube.

Example 10.2: Fluid-filled cylindrical shell
This is a practical example illustrating the use of fluid–structure eigenanalysis to predict the effect of fluid in reducing the resonant frequencies of the structure. Figure 10.3a shows the cross-sectional dimensions of a cylindrical shell filled with water. Since the cylinder is assumed to be infinitely long, the problem is modeled in 2-D to compute the resonant frequencies of the coupled problem. This shell is modeled using two-noded beam finite elements along the circumference and the boundary element nodes of the fluid inside coincide with the finite element nodes of the shell. The computed frequencies along with the analytical frequencies are presented in Table 10.2.

The first fifteen computed eigenmodes of the shell are shown in Figure 10.3c. The analytical frequencies were computed from the frequency expression for the fluid-filled shell given in the reference by Yu [188]. They differ slightly from those reported by Yu because calculations were performed using the British System of units in the reference by Yu, whereas here SI units are used and the resulting rounding of the numbers in the conversion process could be attributed to the difference.

In the boundary element discretization of the fluid, four different cases were investigated to study the effect of putting internal points in the fluid domain. As was pointed out in Section 9.4 [131], the boundary element eigenanalysis invariably needs internal points to ensure that eigenmodes with purely interior nodal lines are not missed. For example, in the case of a circular acoustic fluid domain with hard boundary, eigenmodes that have concentric ring shaped pressure contours will be missed if no internal point is used. In this example, therefore, we solve with and without internal points to confirm this observation in the case of fluid–structure problems. From the results presented in Table 10.2, it is seen that when no internal points are used the breathing mode of the shell, mode 13, is missed and as more internal points are added, the accuracy of the frequencies improves. In order to compare results, finite element–finite element coupled analysis of the problem was performed using the 2-D acoustic fluid elements in the ANSYS program [160]. The finite element mesh used is shown in Figure 10.3b and the computed frequencies are listed in Table 10.2.

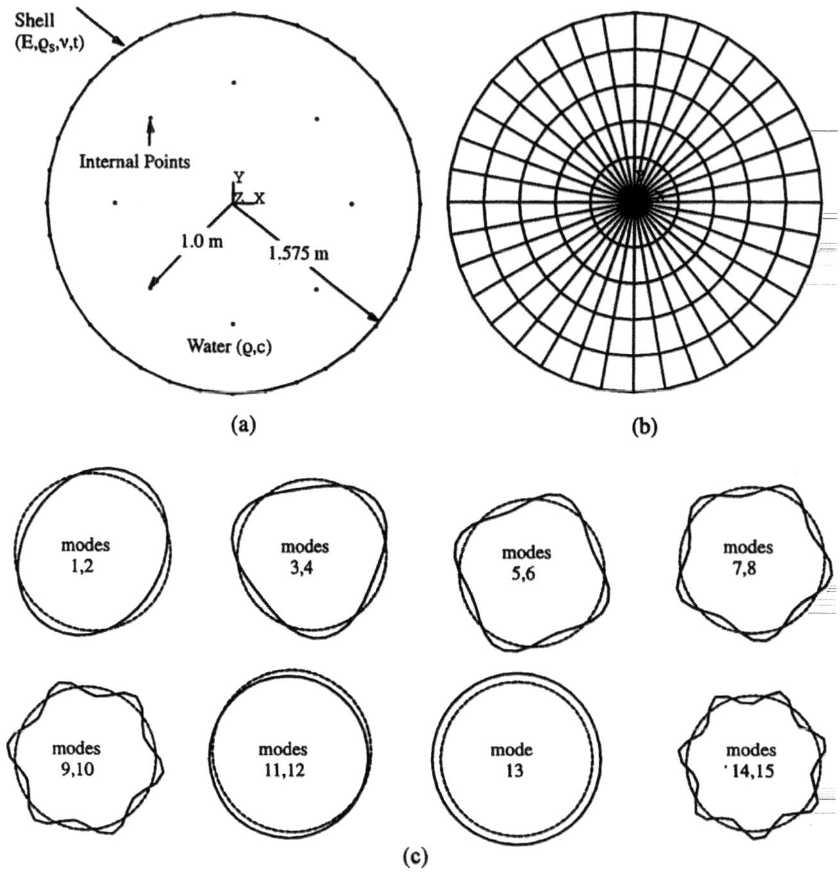

Figure 10.3. Fluid-filled shell model and mode shapes (shell: Young's modulus $E = 2.068 \times 10^{11}$ N/m^2, density $\rho_s = 2403$ kg/m^3, Poisson's ratio $\nu = 0$, thickness $t = 0.102$ m; Water: density $\rho = 1000$ kg/m^3, speed of sound $c = 1500$ m/s). (a) boundary element model; (b) finite element model; (c) shell mode shapes.

Example 10.3: Loudspeaker-box problem

This is a problem chosen to represent a high-fidelity loudspeaker enclosure design application encountered in audio engineering. The problem configuration considered is shown in Figure 10.4, where the loudspeaker-box assembly is modeled in 2-D. The semi-circular shell represents the driver surface and it is fixed on to the box which represents the loudspeaker enclosure, the inside of which is filled with air.

This problem is a representative example of a fluid–structure problem encountered in the acoustical design of a driver-enclosure. Material properties of the box and the shell are chosen to represent a wooden enclosure with a driver made of cloth reinforced plastic, respectively. The property values shown in the Figure 10.4 are taken from reference [189]. Jones et al. [190] point out that a typical material used to construct the driver cone is cobex, which is relatively stiff and light. Since the property values of cobex are not readily available, we have used cloth reinforced plastic which exhibits similar property characteristics.

Table 10.2. Resonant frequencies of the fluid–filled shell (Hz).

	Theory		BE–FE (with different no. of internal points)					
Mode	Free parameter in the freq. expression ($n = 0, 1, 2, \ldots$)	Freq.	0	1	5	9	FE–FE	Freq. of shell w/o fluid inside FE
1, 2	2	25.157	25.158	25.158	25.160	25.159	25.163	47.160
3, 4	3	78.277	79.110	79.105	79.108	79.115	79.176	133.27
5, 6	4	161.10	164.72	164.72	164.73	164.78	165.13	255.23
7, 8	5	274.02	283.66	283.66	283.67	283.69	284.91	412.20
9, 10	6	416.56	436.71	436.71	436.73	436.73	439.45	603.76
11, 12	1	449.64	553.91	553.90	490.58	466.14	445.64	–
13	0	505.59	–	511.57	514.62	515.43	495.06	829.66
14, 15	7	587.41	624.00	624.00	624.00	624.00	628.78	939.81

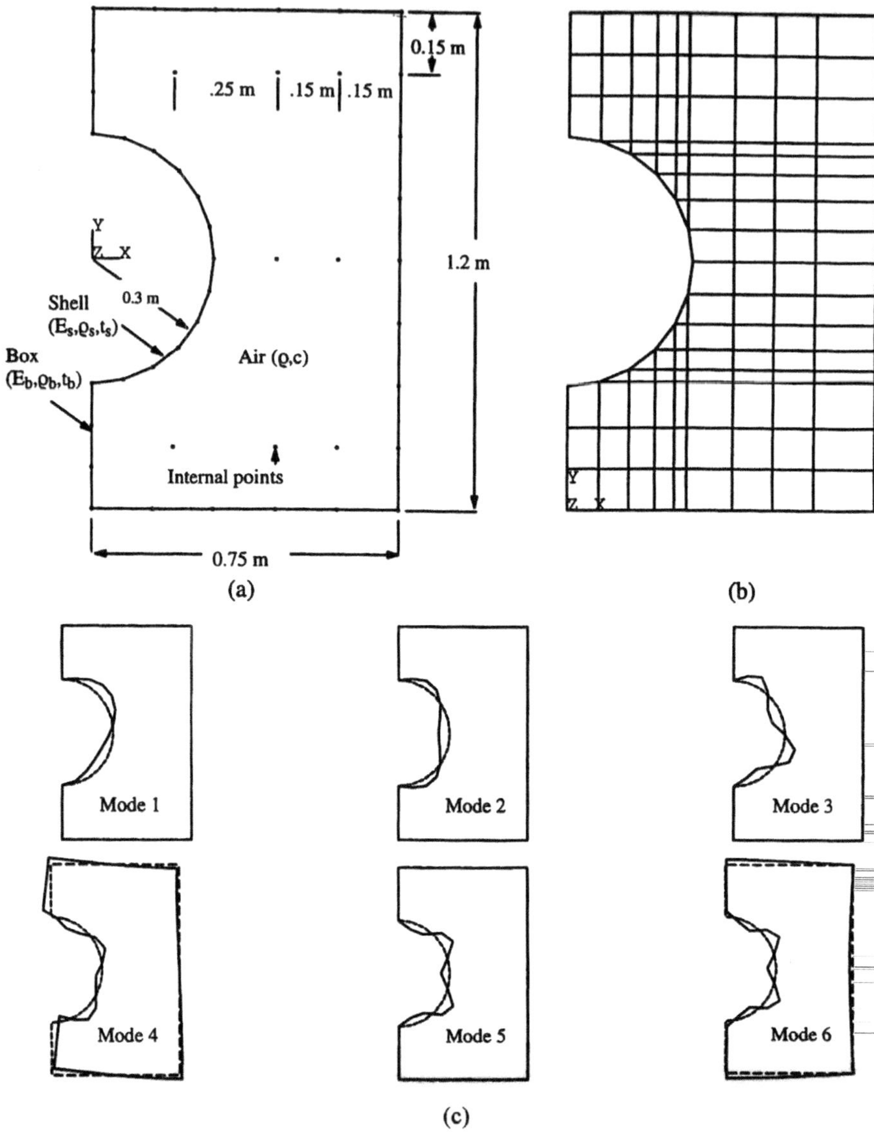

Figure 10.4. Loudspeaker enclosure model and mode shapes (box: Young's modulus $E_b = 1 \times 10^9$ N/m^2, density $\rho_b = 450$ kg/m^3, thickness $t_b = 0.025$ m; Shell: $E_s = 2 \times 10^9$ N/m^2, $\rho_s = 1500$ kg/m^3, $t_s = 0.0005$ m; air: $\rho = 1.12$ kg/m^3, speed of sound $c = 340$ m/s). (a) boundary element model; (b) finite element model; (c) Mode shapes.

The eigenfrequencies of the loudspeaker box assembly are listed in Table 10.3 and the mode shapes are shown in Figure 10.4c. Boundary element–finite element coupled analysis shows good agreement with the finite element–finite element results. In the case of boundary element–finite element coupled analysis, the frequencies computed with and without internal points are quite close to each other indicating that, within the frequency range of the six modes presented, the problem has no modes that have mode shapes with nodal lines purely interior to the boundary. However, for routine

Table 10.3. Resonant frequencies of the loudspeaker-box enclosure (Hz).

	BE–FE			
Mode	Zero internal points	Eight internal points	FE–FE	FE (enclosure without air inside)
1	1.177	1.177	1.178	1.313
2	2.691	2.691	2.692	2.791
3	5.053	5.053	5.056	3.377
4	6.562	6.562	6.562	5.299
5	7.867	7.879	7.875	6.670
6	8.591	8.618	8.623	8.231

computations it is recommended that internal points are always used in order to ensure capturing all of the eigenmodes of a given problem. Also listed in Table 10.3 are the frequencies of the loudspeaker box without air filled inside for comparing the effect of air in the enclosure. It is evident that the air inside the enclosure significantly affects the resonant frequencies of the loudspeaker box assembly.

10.3. Acoustic eigenproblem for enclosures with dissipative boundaries

The damped system resonant modes of acoustic cavities, such as the automobile passenger cabin lined with foam and fibrous type material, is considered in this section. In the previous section we have dealt with acoustic enclosure boundaries which are acoustically hard where there is no loss of energy at the boundary walls and the boundary condition is given by $\partial P/\partial n = 0$. We have also dealt with the case in which the enclosure boundary is a vibrating structure in contact with the acoustic fluid where the boundary condition becomes $\partial P/\partial n = \rho\omega^2 \boldsymbol{u} \cdot \boldsymbol{n}$ causing the energy to be transmitted to the vibrating structure. Another situation is where a part of the enclosure boundary may be completely open to the surrounding atmosphere and in equilibrium with the enclosure acoustic fluid. In this case the boundary condition is simply given by $P = 0$.

When the enclosure boundary is neither acoustically hard nor completely open to the atmosphere, the walls will absorb some energy and the rest will be contained within the acoustic fluid medium. Here, we will present a simple yet effective technique to incorporate a dissipation term in the acoustic eigenvalue formulation to account for the absorption of sound waves at the enclosure boundary walls. The boundary element discretization of acoustic cavities with sound absorption leads to a non-symmetric damped eigenvalue system which is a quadratic eigenvalue problem. A Lanczos based algorithm designed to extract eigenvalues and modes of this non-symmetric quadratic eigenproblem will be presented in Chapter 11. In order to be computationally efficient, the algorithm is designed to avoid the doubling of the system matrices that is normally done in the matrix augmentation process used to linearize the quadratic eigenvalue problem.

A brief outline of the formulation is as follows: First, we will restate the acoustic boundary element eigenproblem equation without the dissipation term, as derived in Chapters 8 and 9. Then, a simple method of incorporating boundary absorption into the acoustic wave equation discretization process will be shown. We will, next, show how to incorporate the absorption term in the boundary element matrix equation. Finally, we will present several example problems, solved using the Lanczos algorithm for non-symmetric quadratic eigenproblem. The Lanczos eigenvalue solver itself will be described Chapter 11.

10.3.1. Acoustic boundary element eigenproblem

As shown in earlier chapters, the acoustic wave equation governing the acoustic pressure p in a fluid is given by:

$$\nabla^2 p = \frac{1}{c^2}\partial^2 p/\partial t^2 \tag{10.9}$$

c is the speed of sound and t is the time. For the harmonic oscillations of pressure in the sound pressure wave problem considered, $p = P\mathrm{e}^{\mathrm{j}\omega t}$. P is the pressure amplitude, ω is the circular frequency and $j = \sqrt{-1}$. Substituting for p in equation (10.9), the Helmholtz equation governing the amplitude of pressure oscillations is obtained.

$$\nabla^2 P = -(\omega/c)^2 P \tag{10.10}$$

The boundary element algebraic eigenvalue problem for this governing equation is given by [see eqns. (8.16), (8.52) and (10.2)]:

$$[G]\{\partial P/\partial n\} - [H]\{P\} = (\omega/c)^2[M]\{\Phi\} \tag{10.11}$$

where the mass-type matrix $[M] = ([G][T] - [H][D])$. This matrix equation represents the boundary element discretized, loss-less Helmholtz equation (10.10). In practical applications, however, the acoustic sound pressure partly dissipates as it reflects off of the bounding surfaces of an enclosure such as the carpeted flooring of a room. In the following section, a method of incorporating the dissipation term in the discretized Helmholtz equation (10.11) is described.

10.3.2. Sound absorption at the boundary

Some amount of energy is almost always dissipated due to the absorption of sound at the boundaries of an acoustic domain. Especially, when different sound-absorbing materials are used in the acoustic enclosure walls to deliberately dampen the sound, as shown in Figure 10.5, a way of modeling the sound absorption is needed. In order to account for the dissipated energy, the reference by Craggs [191] shows a damped form of the Helmholtz equation (10.10). For the absorption of sound at the boundaries of an acoustic cavity, a similar approach is used in the linear momentum equation that relates the fluid pressure gradient and the velocity at the boundary:

$$\partial p/\partial n = -\rho(\partial \boldsymbol{v}/\partial t) \cdot \boldsymbol{n} \tag{10.12}$$

where ρ is the fluid density, $\boldsymbol{v}$ is the velocity vector, and $\boldsymbol{n}$ is the outward normal. To account for the energy dissipated at the boundaries, an absorption term is introduced on the right-hand side of equation (10.12):

$$\partial p/\partial n = -\rho(\partial \boldsymbol{v}/\partial t) \cdot \boldsymbol{n} + R\nabla \cdot \boldsymbol{v} \tag{10.13}$$

In the absorption term, the reason for using the velocity divergence $\nabla \cdot \boldsymbol{v}$, as opposed to $\boldsymbol{v}$ as in Cragg's paper, comes from the consideration of the consistency of units, where the acoustic resistance R is in rayl (N s/m^3). In addition, Zienkiewicz and Newton [186] give the relationship $\partial p/\partial n = -(1/c)\ \partial p/\partial t$ to account for the energy loss due to the

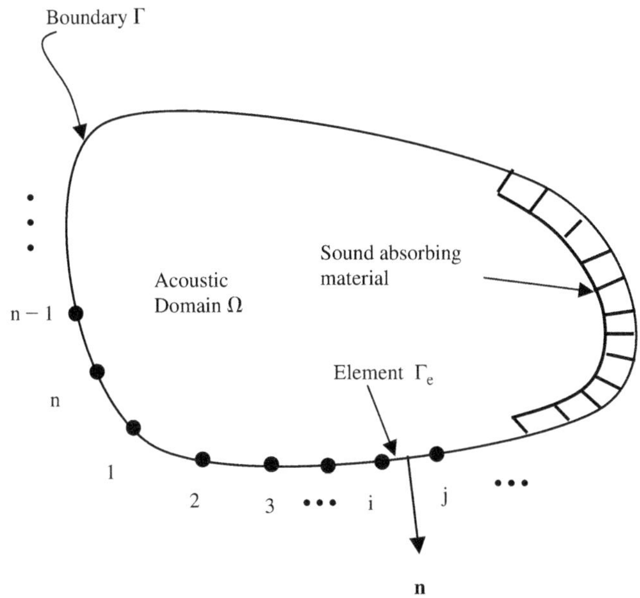

Figure 10.5. Acoustic cavity.

outbound pressure waves at an open boundary for plane wave situations. Since $\partial p/\partial t$ is related to $\nabla \cdot \boldsymbol{v}$ via the conservation of mass equation:

$$\nabla \cdot \boldsymbol{v} = -\frac{1}{\rho c^2}\frac{\partial p}{\partial t} \tag{10.14}$$

of a compressible fluid, the boundary absorption term used in equation (10.13) would be appropriate. Substituting for $\nabla \cdot \boldsymbol{v}$ from equation (10.14) into equation (10.13) the pressure gradient at an absorbing boundary is written as:

$$\partial p/\partial n = -\rho \partial \boldsymbol{v}/\partial t \cdot \boldsymbol{n} - \beta\left(\frac{1}{c}\frac{\partial p}{\partial t}\right) \tag{10.15}$$

$\beta = R/\rho c$ is the non-dimensional absorption coefficient of the sound absorbing material at the boundary. The acoustic resistance R for foam and fibrous type material is given as a function of frequency by Craggs [191].

For the harmonic oscillations in acoustical problems, the boundary pressure and velocity in equation (10.15) are represented in complex exponential form as $p = P\mathrm{e}^{\mathrm{j}\omega t}$ and $v = V\mathrm{e}^{\mathrm{j}\omega t}$, respectively. We are not using the boundary velocity here since the fluid–structure interface boundary is assumed to be stationary. The boundary pressure gradient at a sound absorbing boundary is, then, obtained from equation (10.15) as:

$$\partial P/\partial n = -\mathrm{j}(\omega/c)\beta P \tag{10.16}$$

At the absorbing boundary, however, the particle velocity normal to the surface may not be zero, since there exists a non-zero pressure gradient when $\beta \neq 0$. Therefore, the absorbing boundary can be described as "quasi-rigid" even though the boundary itself is not oscillating as in a fluid–structure coupling interface.

Often in acoustical modeling, the impedance of an absorbing boundary is given. So, a relationship between the absorption coefficient β and the acoustic impedance Z at the absorbing boundary can be obtained if the particle velocity v_β normal to the surface and the pressure are known. For the harmonically varying acoustic waves, $v_\beta = V_\beta e^{j(\omega t+\alpha)}$, where V_β is the velocity magnitude and α is the phase angle of the velocity with respect to the pressure at the absorbing boundary. Using the relationship between the pressure gradient and the particle velocity normal to the boundary, the expression for V_β can be found as $V_\beta = j\,(\partial P/\partial n)\,e^{-j\alpha}/(\rho\omega)$. Solving for the pressure P from equation (10.16), the impedance Z at the absorbing boundary is found by dividing P by V_β. The resulting expression for the impedance of the absorbing boundary is $Z = (\rho c/\beta)\,e^{j\alpha}$. It is evident from this relationship that the acoustic impedance of the absorbing boundary is complex-valued even though the absorption coefficient β is real-valued. Therefore, in situations where the impedance of a boundary is known, the absorption coefficient to be applied can be calculated from this expression as $\beta = \rho c/|Z|$.

The valid range of the absorption coefficient β is from zero to unity. When $\beta = 0$, there is no energy loss at the boundary and the wall is rigid. For $\beta = 1$, the entire energy incident on the boundary is lost, which represents a boundary left open. For values of β that lie between zero and unity, part of the incident energy on the boundary is lost and the rest is reflected back. Since β is non-dimensional, the question of whether its value can be greater than unity arises. In such a case $R > \rho c$, which implies that the absorbing boundary assists in taking energy out of the system. A plausible explanation can be given for a case where the acoustic medium beyond the absorbing boundary has a characteristic impedance greater than ρc.

10.3.3. Absorption term in the boundary element eigenvalue matrix equation

Using the relationship in equation (10.16), the absorption in an acoustic cavity can be incorporated in the discretized boundary element equation (10.11). Equation (10.16) is rewritten in terms of the nodal pressure gradients as follows:

$$\{\partial P/\partial n\} = -j(\omega/c)\beta\{P\} \tag{10.17}$$

When absorption material is used on a part of the boundary Γ_1, equation (10.11) can be written in partitioned form as:

$$\begin{bmatrix} G_{11} & G_{12} \\ G_{21} & G_{22} \end{bmatrix} \begin{Bmatrix} Q_1 \\ Q_2 \end{Bmatrix} - [H]\{P\} = \left(\frac{\omega}{c}\right)^2 [M]\{\Phi\} \tag{10.18}$$

$\{Q_1\}$ and $\{Q_2\}$ denote pressure gradients at the absorbing and non-absorbing boundary nodes, respectively. Then, substituting from equation (10.17) into equation (10.18), one obtains:

$$\begin{bmatrix} -j\left(\frac{\omega}{c}\right)\beta G_{11} & G_{12} \\ -j\left(\frac{\omega}{c}\right)\beta G_{21} & G_{22} \end{bmatrix} \begin{Bmatrix} P_1 \\ Q_2 \end{Bmatrix} - [H]\{P\} = \left(\frac{\omega}{c}\right)^2 [M]\{\Phi\} \tag{10.19}$$

For the part of the boundary where there is no absorption, we assume acoustically hard boundary condition $\{Q_2\} = 0$. Partitioning the matrix $[C]$ of equation (8.17) as in equation (10.18), we substitute for $\{P_1\} = [C_{11}]\{\Phi_1\} + [C_{12}]\{\Phi_2\}$ in equation (10.19):

$$-j\left(\frac{\omega}{c}\right)\beta \begin{bmatrix} G_{11}C_{11} & G_{11}C_{12} \\ G_{21}C_{11} & G_{21}C_{12} \end{bmatrix} \begin{Bmatrix} \Phi_1 \\ \Phi_2 \end{Bmatrix} - [H][C]\{\Phi\} = \left(\frac{\omega}{c}\right)^2 [M]\{\Phi\} \tag{10.20}$$

Taking the coefficient $-\beta$ into the matrix, and denoting the absorption matrix as $[\overline{C}]$ and $[K]=-[H][C]$, equation (10.20) is rewritten as:

$$j(\omega/c)[\overline{C}]\{\Phi\}+[K]\{\Phi\}=(\omega/c)^2[M]\{\Phi\} \qquad (10.21)$$

Equation (10.21) is the eigenvalue problem for the acoustic cavity with sound absorption. Defining the eigenvalues as $\lambda_i=j(\omega/c)$, and the eigenvectors $\{x_i\}=\{\Phi\}$, the damped system eigenproblem to be solved is:

$$[K]\{x_i\}+\lambda_i[\overline{C}]\{x_i\}+\lambda_i^2[M]\{x_i\}=0 \qquad (10.22)$$

Equation (10.22) represents the generalized quadratic eigenvalue problem and, in general, will yield complex eigenvalues and eigenvectors. The computational approach to solve this eigenvalue problem employing the Lanczos recursion scheme will be shown in the next chapter.

10.4. Examples of acoustic eigenproblem with sound absorption

We will consider two different acoustical cavities to demonstrate the effect of incorporating sound absorption in the acoustic boundary element eigenproblem. One is a long impedance tube and the other is a square cavity. The boundary absorption method formulated in this chapter is also employed in the acoustic finite elements in the ANSYS® general purpose program [160]. So, we compare the boundary element results with those of the finite element eigensolution results from the ANSYS program in cases where a theoretical solution is unavailable. These examples, however, are not designed to quantitatively evaluate the absorption model that was shown in the previous section.

Example 10.4
A long impedance tube of length to width ratio of 20, shown in Figure 10.6, is considered first. The tube wall at $x=L$ is lined with a damping material of absorption coefficient β. Since the aspect ratio is large, the first few undamped mode frequencies for this problem are given by the 1-D frequency expression $f_i=ic/(2L), i=1,2,\ldots,\infty$. Table 10.4 shows the computed frequencies for three different values of absorption coefficient, $\beta=0$, 0.1 and 0.9, along with the undamped theoretical frequencies. The acoustical resonant frequencies are the positive imaginary parts of the conjugate pair of eigenvalues computed.

For the undamped case, the frequencies computed by the BEM are in good agreement with the theoretical values. For the damped case, boundary element frequencies are

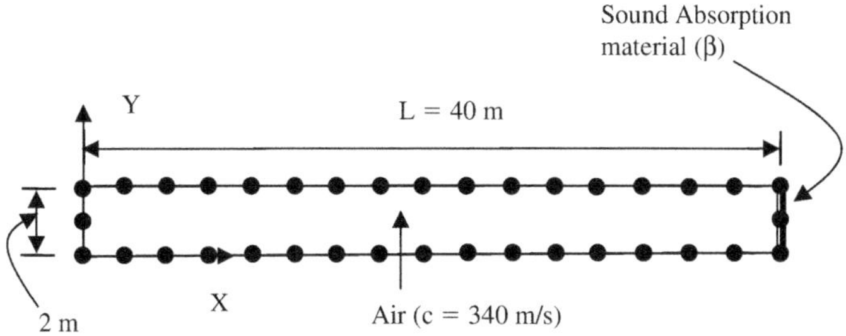

Figure 10.6. Impedance tube with sound absorption (coefficients used: $\beta=0$, 0.1 and 0.9).

Table 10.4. Acoustic resonant frequencies of impedance tube (Hz).

Absorption coefficient (β)	Mode	Theory	BE		FE	
0	1	–	−0.52727113E−05	−0.71259621E−07j	0.13906660E−05	−0.12434703E−04j
	2	0 + 0j	0.52726372E−05	+0.71252287E−07j	0.13872865E−05	+0.12435437E−04j
	3	–	0.00000000E+00	−4.2638284j	0.00000000E+00	−4.2568304j
	4	0 + 4.25j	0.00000000E+00	+4.2638284j	0.00000000E+00	+4.2568304j
	5	–	0.00000000E+00	−8.5312338j	0.00000000E+00	−8.5547171j
	6	0 + 8.50j	0.00000000E+00	+8.5312338j	0.00000000E+00	+8.5547171j
	7	–	0.00000000E+00	−12.905870j	0.00000000E+00	−12.935041j
	8	0 + 12.75j	0.00000000E+00	+12.905870j	0.00000000E+00	+12.935041j
	9	–	0.00000000E+00	−17.295136j	0.00000000E+00	−17.439604j
	10	0 + 17.00j	0.00000000E+00	+17.295136j	0.00000000E+00	+17.439604j
0.1	1		0.15200149E−08	−0.56715029E−09j	0.16049927E−09	−0.73970017E−11j
	2	−0.136 + 0j	−0.13493624	+0.00000000E+00j	−0.13573492	+0.00000000E00j
	3		−0.13690670	−4.2638323j	−0.13661136	−4.2567951j
	4		−0.13690670	+4.2638323j	−0.13661136	+4.2567951j
	5		−0.13819908	−8.5310962j	−0.13927476	−8.5546444j
	6		−0.13819901	+8.5310962j	−0.13927476	+8.5546444j
	7		−0.14306932	−12.905832j	−0.14382862	−12.934927j
	8		−0.14306936	+12.905832j	−0.14382862	+12.934927j
	9		−0.14803122	−17.294763j	−0.15044929	−17.439440j
	10		−0.14803126	+17.294763j	−0.15044929	+17.439440j
0.9	1		−0.74389602E−08	+0.53838077E−07j	0.46269145E−08	+0.49989396E−08j
	2	−1.992 + 0j	−1.9627462	+0.00000000E+00j	−1.9886941	−0.57453587E−07j
	3		−2.0131707	−4.2577669j	−2.0084570	−4.2424017j
	4		−2.0131707	+4.2577669j	−2.0084570	+4.2424017j
	5		−2.0478985	−8.4778993j	−2.0699368	−8.5233054j
	6		−2.0478985	+8.4778994j	−2.0699370	+8.5233054j
	7		−2.1870322	−12.858083j	−2.1805021	−12.880197j
	8		−2.1870323	+12.858083j	−2.1805023	+12.880197j
	9		−2.3191496	−17.111469j	−2.3556919	−17.346699j
	10		−2.3191497	+17.111469j	−2.3556920	+17.346699j

compared with finite element results since the theoretical expression is not readily available. However, for the non-oscillatory mode, mode 2, a theoretical formula can be obtained from the following frequency equation of this 1-D acoustic problem with sound absorption at $x = L$:

$$\tanh(rL) = -(\beta s)/(cr) \tag{10.23}$$

where $r = \alpha + jk$ and $s = \sigma + j\omega$ are the complex wave number and frequency, respectively. This expression is similar to the one found in the reference by Kinsler et al. [187], where the boundary absorption is dealt with under the heading of reverberation of normal modes. The specific expression referred to here is equation (13.43) in reference [187]. Setting $\omega = 0$ in equation (10.23), we obtain the eigenvalue of the non-oscillatory mode:

$$\sigma = (c/L)\tanh^{-1}\beta \tag{10.24}$$

The theoretical frequencies of mode 2 calculated using equation (10.24) are listed in Table 10.4 for comparison with the computed results; the agreement is good. For all other modes, the boundary element frequencies are compared with the finite element frequencies. The close agreement between the two indicates the validity of the numerical approach presented for the incorporation of sound absorption term in the boundary element eigenproblem.

Example 10.5

A square cavity, shown in Figure 10.7, is considered next. The absorption free eigenfrequencies for this problem are given by:

$$f_i = (c/2)[(l_x/L_x)^2 + (l_y/L_y)^2]^{1/2}, \quad l_x, l_y = 0, 1, 2, \ldots, \infty \tag{10.25}$$

where L_x and L_y are the length and width of the domain, respectively. The absorption coefficients at the absorption layer and at the walls β_a and β_b, respectively, are set to zero so that the computed frequencies can be compared with the theoretical frequencies given by the expression in equation (10.25). For the boundary element discretization,

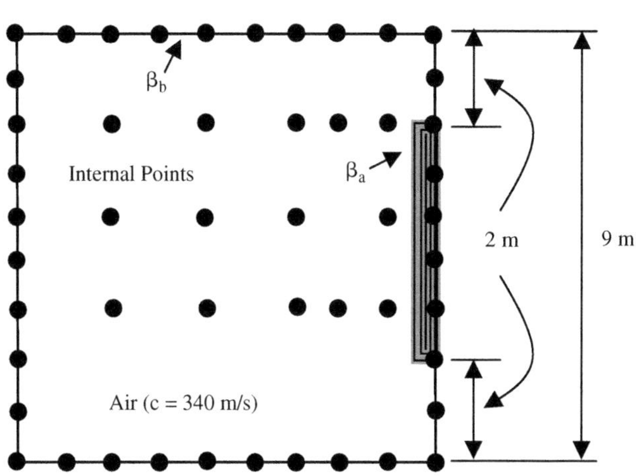

Figure 10.7. Boundary element mesh of square cavity (absorption coefficient: β_a = at the absorption layer, β_b = at the walls).

Table 10.5. Acoustic resonance frequencies of cavity with zero absorption ($\beta_a = \beta_b = 0$).

No.	Theoretical frequencies (Hz)	BE		FE	
1	–	0.90094705E−05	−0.16054364E−05j	−0.73030743E−04	−0.98103609E−06j
2	0	−0.90096915E−05	0.16051918E−05j	0.73051942E−04	0.98157164E−06j
3	–	0.00000000E+00	−19.031585j	0.00000000E+00	−18.984927j
4	18.8889	0.00000000E+00	19.031585j	0.23827455E−06	18.984927j
5	–	−0.19916942E−06	−19.127708j	0.29067474E−06	−18.984928j
6	18.8889	0.00000000E+00	19.127708j	0.29076915E−06	18.984928j
7	–	0.00000000E+00	−26.972142j	0.00000000E+00	−26.848742j
8	26.7129	0.00000000E+00	26.972142j	0.00000000E+00	26.848742j
9	–	0.00000000E+00	−39.214077j	0.00000000E+00	−38.548881j
10	37.7778	0.00000000E+00	39.214077j	0.00000000E+00	38.548881j

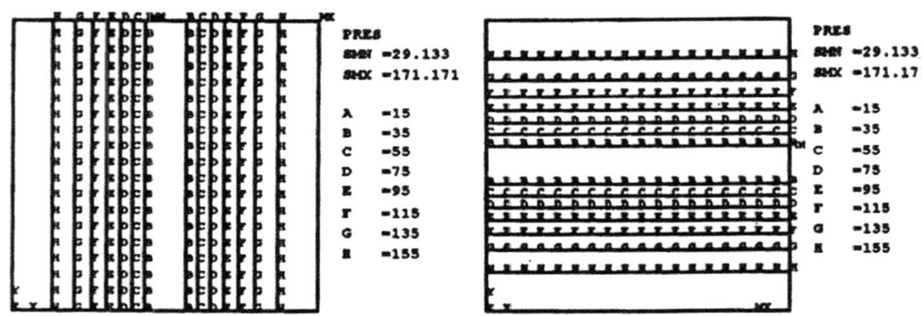

(Mode 4) (Mode 6)

Figure 10.8. Square cavity mode shapes for zero absorption ($\beta_a = \beta_b = 0$).

Table 10.6. Acoustic resonance frequencies of cavity with uniform absorption ($\beta_a = \beta_b = 0.2$).

No.	BE		FE	
1	−0.69521872E−10	−0.32956126E−09j	0.53712161E−09	0.51497840E−09j
2	−4.8788198	0.00000000E+00j	−4.9431474	0.00000000E+00j
3	−3.8496925	−18.905690j	−3.7763896	−18.905246j
4	−3.8496920	18.905690j	−3.7763896	18.905246j
5	−3.9248480	−18.973677j	−3.7763897	−18.905247j
6	−3.9248480	18.973678j	−3.7763897	18.905247j
7	−5.3478454	−26.783942j	−5.1516059	−26.823973j
8	−5.3478454	26.783942j	−5.1516060	26.823973j
9	−4.5679968	−39.211240j	−3.9654993	−38.695024j
10	−4.5679969	39.211240j	−3.9654994	38.695024j

Table 10.7. Acoustic resonance frequencies of cavity with non-uniform absorption ($\beta_a = 0.9$, $\beta_b = 0$).

No.	BE		FE	
1	−0.32277792E−07	−0.31245974E–07j	0.70742767E−08	−0.37968327E−08j
2	−3.3779687	0.0000000E+00j	−3.4207326	0.0000000E+00j
3	−0.61991593	−19.372290j	−0.61317013	−19.236889j
4	−0.61991592	19.372290j	−0.61317014	19.236889j
5	−3.3924172	−19.793552j	−3.2898905	−19.755583j
6	−3.3924172	19.793552j	−3.2898902	19.755583j
7	−1.3665337	−27.344634j	−1.3564097	−27.252889j
8	−1.3665337	27.344634j	−1.3564100	27.252889j
9	−0.39190014	−39.427160j	−0.32248800	−38.721521j
10	−0.39190016	39.427160j	−0.32248844	38.721522j

36 boundary nodes and 14 internal nodes are used. The need for internal nodes to improve the accuracy of computed frequencies was pointed out in Chapter 9. The frequencies shown in Table 10.5 include the theoretical results and a 10×10 mesh finite element [160] results for comparison with the boundary element solution. The acoustic resonant frequencies are the positive imaginary parts of the conjugate pair of eigenvalues computed. The five eigenfrequencies from the BEM compare favorably with the finite element solution and these are also close to the theoretical frequencies. The mode shapes 4 and 6 of the cavity pressure amplitude for this absorption-free case are shown in Figure 10.8.

Example 10.6
The square cavity problem is solved with uniform boundary absorption by taking $\beta_a = \beta_b = 0.2$. Table 10.6 shows the frequencies computed by the BEMs and FEMs using the same discretization as in example 10.5. The eigenvalues $\lambda_i = \sigma_i \pm j\omega_i$ now have a non-zero real part. The imaginary part gives the damped resonant frequency $f_i = \omega_i/(2\pi)$. Compared to the non-absorption case, the frequencies have slightly reduced for modes 3–8. Mode 2 shows a non-oscillatory exponential decay response. The boundary element results compare well with the finite element results at the lower modes. With an increasing number of internal points, the boundary element results will improve. The mode shapes 2, 4 and 6 obtained from the finite element results are plotted in Figure 10.9.

Example 10.7
Here the absorption is non-uniform, $\beta_a = 0.9$ and $\beta_b = 0$. Table 10.7 shows the frequency results. Modes 3–10 show an increase from the undamped frequencies. Mode 2 again shows a non-oscillating decay. The plots of mode shapes 2, 4 and 6 obtained from the finite element analysis are shown in Figure 10.10. It is interesting to note the concentration of contours near the absorption layer in these plots.

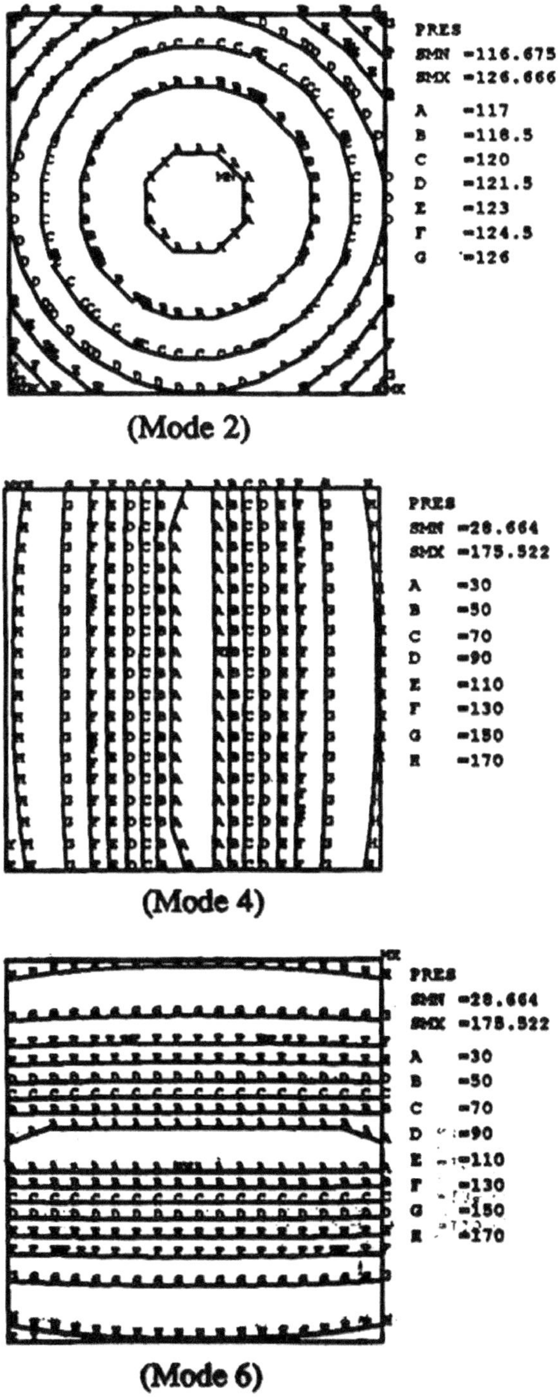

Figure 10.9. Square cavity mode shapes for uniform absorption ($\beta_a = \beta_b = 0.2$).

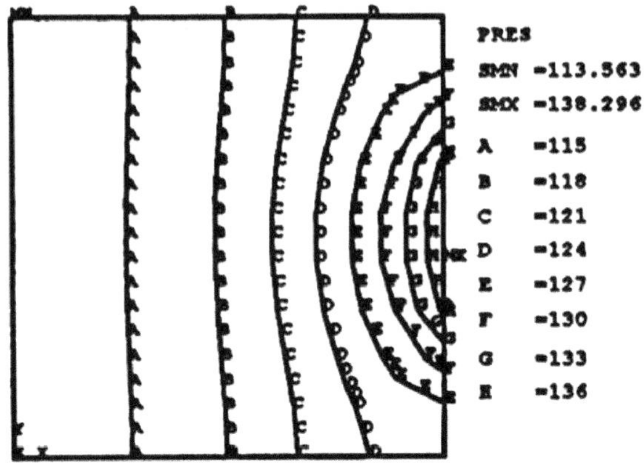

(Mode 2)

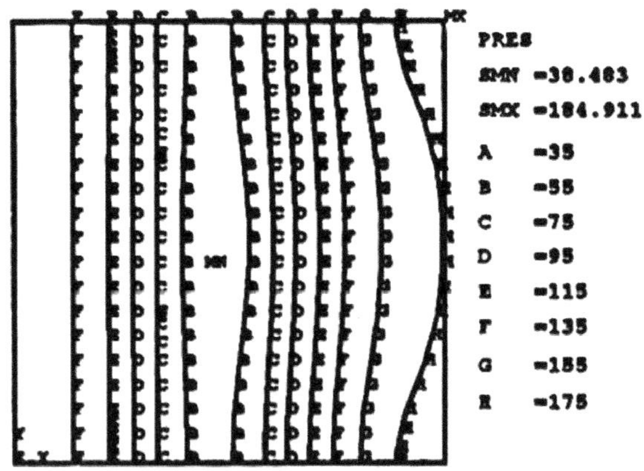

(Mode 4)

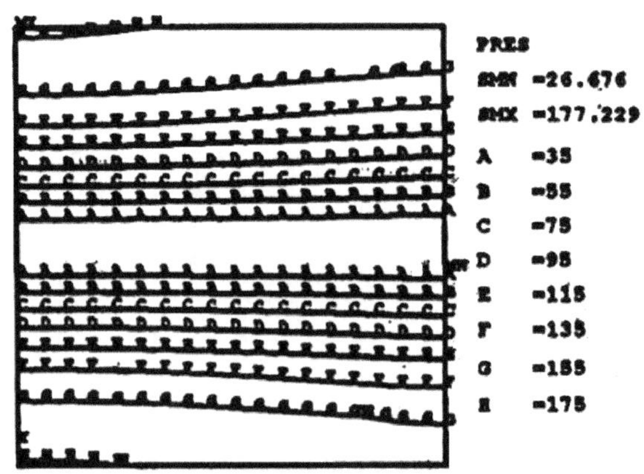

(Mode 6)

Figure 10.10. Square cavity mode shapes for non-uniform absorption ($\beta_a = 0.9$ and $\beta_b = 0$).

Chapter 11

Solution Methods of Eigenvalue Problems

11.1. Introduction

In Chapters 1 through 10, we have described in detail the boundary element discretization process and, specifically, the formulation of the boundary element eigenvalue problem encountered in elastodynamics, and acoustic fluid-structure analyses. Extraction of the eigenvalues and mode shapes from these eigenvalue problems, often, is a computationally intensive process owing to the fact that most practical applications involve large matrices. Typical applications might be, finding the resonant frequencies and mode shapes of mechanical systems like building structures, machinery parts, acoustic enclosures such as a concert hall, or acoustic fluid–structure systems such as a submarine under water. Therefore, a computationally efficient eigenvalue extraction procedure is crucial to the success of the overall analysis approach using the boundary element method (BEM).

The finite element based eigenvalue problems, usually, result in symmetric matrices. While the boundary element eigenvalue problem matrices are non-symmetric, they represent only the boundary discretization and are smaller in size than the symmetric finite element matrices obtained by domain discretization. Therefore, an efficient eigenvalue extraction procedure for the non-symmetric generalized eigenvalue problem would compliment the benefit gained by the smaller size boundary element matrices.

11.2. Lanczos-based subspace approach

In most computational situations, we are interested in finding the lowest few eigenvalues and mode shapes of large matrices. This is true for practical structural dynamic applications where the lowest and mid range resonant frequencies are of interest in the design of structural components and assemblies. The Lanczos method of computing eigenvalues fits quite well for this situation, as the method has the ability to extract a few of the eigenvalues of large matrices, say 50 eigenvalues and mode shapes of a large matrix of size 100,000 or more. In acoustic fluid–structure problems, it is often required to search for the resonant frequencies of the coupled system in the lowest and the middle ranges of the frequency spectrum of the eigenproblem since the fluid domain resonant frequencies are closely spaced. The Lanczos eigenvalue extraction method has been quite successfully employed in finite element based eigenvalue problems where the system matrices are symmetric. Since the boundary eigenvalue problem matrices are non-symmetric, in this chapter we will show the Lanczos eigenvalue schemes designed specifically for such systems.

Two widely used eigenvalue computational methods are the Subspace Iteration and the Lanczos Recursion methods. Both of these methods reduce a large generalized eigenproblem shown below

$$[K]\{x_i\} = \lambda_i[M]\{x_i\} \quad i = 1, 2, \ldots, n \tag{11.1}$$

into smaller-sized standard eigenproblem through orthogonal transformation into a subspace. The eigenvalues, $\mu_j, j = 1, 2, \ldots m$ of this transformed reduced size eigenvalue problem

$$[\overline{K}]\{q_j\} = \mu_j\{q_j\} \quad j = 1, 2, \ldots, m \quad (m < n) \tag{11.2}$$

approximate the eigenvalues λ_i of the original eigenvalue problem. These eigenvalues converge to that of the original problem as the subspace size m is progressively increased.

Owing to the superior convergence properties, the Lanczos recursion method is quite widely employed in solving large eigenvalue problems arising in finite element structural dynamic problems. Therefore, in this chapter the Lanczos recursion method will be presented for the solution of the boundary element eigenvalue problems.

11.3. Lanczos recursion method

The boundary element eigenvalue problems that we formulated in this book in Chapters 7 through 10 deal with non-symmetric generalized eigenvalue problems. The eigenvalue problem presented in equation (8.22) is rewritten here for consideration.

$$[A]\{x_i\} = \lambda_i[B]\{x_i\} \tag{11.3}$$

where:

$\{x_i\}$ are the eigenvectors $\lfloor P_i, \partial P_i/\partial n \rfloor$;
λ_i are the eigenvalues which represent the square of the resonant wave numbers k_i in acoustic problems ($\lambda_i = k_i^2 = \omega_i/c$).

Matrices $[A]$ and $[B]$ in the boundary element eigenvalue problem are non-symmetric. To solve a non-symmetric generalized eigenvalue problem, two possible approaches can be used to transform the eigenproblem into a subspace. These are the Arnoldi's method, and the Lanczos method. Both of these methods are based on the common origin of Krylov subspace techniques. In this book, however, the Lanczos algorithm will be described for the non-symmetric eigenvalue problem.

Solution of the non-symmetric eigenvalue problem using the Lanczos algorithm involves a two-sided recursion scheme applied to transform the generalized eigenvalue problem given by equation (11.3) into a standard eigenvalue problem. In reference [192] detailed formulation of the Lanczos recursion is presented which will be used here to concisely present the method. The two sets of Lanczos vectors generated by the recursion form the biorthogonal bases for the subspace onto which the generalized eigenproblem is projected to yield the standard eigenvalue problem. The matrix of the standard eigenvalue problem in the Krylov subspace results in a tri-diagonal form lending itself for easy and efficient extraction of its eigenvalues μ_j. As the subspace size m is increased, μ_j converge to the eigenvalues λ_i of the original eigenproblem. In the limit when $m = n$, $\lambda_i = \mu_j$.

11.3.1. Lanczos recursion for the standard eigenvalue problem

In order to develop the Lanczos scheme for the generalized eigenvalue problem, the two-sided Lanczos recursion as applied to the following standard eigenvalue problem will be reviewed briefly.

$$[A]\{x_i\} = \lambda_i\{x_i\} \tag{11.4}$$

The two-sided recursion is applied to the matrix $[A]$ and its transpose $[A]^{\mathrm{T}}$ simultaneously as described in reference [192]. Starting from two arbitrarily chosen vectors $\{v_1\}$ and $\{w_1\}$, two sets of Lanczos vectors $\{v_j\} = \}v_1, v_2, \ldots v_m\}$ and $\{w_j\} = \{w_1, w_2, \ldots w_m\}$ are generated from the right and left Krylov sequence of vectors. The right and left Krylov sequence of vectors are given by

$$\{v_1, [A]v_1, [A]^2 v_1, \ldots, [A]^{m-1} v_1\} \tag{11.5a}$$

and

$$\{w_1, ([A]^{\mathrm{T}})w_1, ([A]^{\mathrm{T}})^2 w_1, \ldots, ([A]^{\mathrm{T}})^{m-1} w_1\} \tag{11.5b}$$

respectively. The set of m right and left Lanczos vectors $\{v_j\}$ and $\{w_j\}$ form the biorthonormal bases for the subspace onto which the given matrix $[A]$ is projected. As pointed out in references [192] and [193], a non-symmetric matrix $[A]$ has two sets of eigenvectors $\{x_i\}$ and $\{z_i\}$ which are the right and left eigenvectors belonging to the original problem in equation (11.4) and the transposed eigenproblem

$$[A]^{\mathrm{T}}\{z_i\} = \lambda_i\{z_i\} \tag{11.6}$$

The eigenvector matrices $[X] = [x_1, x_2, \ldots, x_n]$ and $[Z] = [z_1, z_2, \ldots, z_n]$ will be biorthonormal yielding

$$[Z]^{\mathrm{T}}[X] = [I] \tag{11.7}$$

where $[I]$ is the identity matrix. The Lanczos scheme for the standard eigenvalue problem, therefore, is devised to generate two sets of vectors $[V] = [v_1, v_2, \ldots, v_m]$ and $[W] = [w_1, w_2, \ldots, w_m]$ from the right and left Krylov sequence such that

$$[W]^{\mathrm{T}}[V] = [I] \tag{11.8}$$

$[V]$ and $[W]$ are biorthonormal. The two-sided Lanczos recursion used to generate the Lanczos vectors $\{v_j\}$ and $\{w_j\}$ is given in references [178, 192, 193]. Here we present the two-sided recursion for the standard eigenvalue problem [194] as an introduction to formulating the recursion scheme for the generalized eigenvalue problem posed in equation (11.3).

For $j = 1, 2, \ldots, m\,(m \le n)$ choose $\{v_1\}$ and $\{w_1\}$ such that $\{w_1\}^{\mathrm{T}}\{v_1\} = 1$.

$$\{\overline{v}_{j+1}\} = [A]\{v_j\} - \alpha_j\{v_j\} - \beta_j\{v_{j-1}\} \tag{11.9a}$$

$$\{\overline{w}_{j+1}\} = [A]^{\mathrm{T}}\{w_j\} - \alpha_j\{w_j\} - \delta_j\{w_{j-1}\} \tag{11.9b}$$

When $j = 1$, $\beta_1\{v_0\} = 0$, and $\delta_1\{w_0\} = 0$.

$$\alpha_j = \{w_j\}^{\mathrm{T}}[A]\{v_j\} \tag{11.10}$$

$$\delta_{j+1} = |\{\overline{w}_{j+1}\}^{\mathrm{T}}\{\overline{v}_{j+1}\}|^{1/2} \tag{11.11}$$

$$\beta_{j+1} = \delta_{j+1}\mathrm{sign}(\{\overline{w}_{j+1}\}^{\mathrm{T}}\{\overline{v}_{j+1}\}) \tag{11.12}$$

If $\delta_{j+1} = 0$, restart the recursion with new $\{v_1\}$ and $\{w_1\}$.

$$\{v_{j+1}\} = \frac{\{\overline{v}_{j+1}\}}{\delta_{j+1}} \tag{11.13}$$

$$\{w_{j+1}\} = \frac{\{\overline{w}_{j+1}\}}{\beta_{j+1}} \tag{11.14}$$

The set of m Lanczos vectors generated by this recursion provides a biorthogonal transformation of the matrix $[A]$ yielding a tridiagonal matrix of size m given by

$$[T] = [W]^{\mathrm{T}}[A][V] \tag{11.15}$$

This tridiagonal matrix can be directly formed using the scalar coefficients generated in the Lanczos recursion steps presented in equations (11.9) through (11.14) as follows:

$$[T] = \begin{bmatrix} \alpha_1 & \beta_2 & & & & \\ \delta_2 & \alpha_2 & \beta_3 & & & \\ & \delta_3 & \alpha_3 & \cdot & & \\ & & \cdot & \cdot & \cdot & \\ & & & \cdot & \cdot & \cdot \\ & & & & \cdot & \cdot & \beta_m \\ & & & & & \delta_m & \alpha_m \end{bmatrix} \tag{11.16}$$

The m eigenvalues of the tridiagonal matrix $[T]$ approximate m eigenvalues of the matrix $[A]$ at one end of its eigenvalue spectrum. Theoretically, when $m = n$, all of the eigenvalues of $[A]$ would be found. The proof of the biorthogonal transformation and issues relating to the implementation details will be presented in the next section where we present the recursion for the generalized non-symmetric eigenvalue problem that we started with in equation (11.3).

11.3.2. Lanczos algorithm for the generalized eigenvalue problem

The application of a two-sided Lanczos recursion to the generalized eigenvalue problem in equation (11.3) requires the generalization of the recursion presented for the standard eigenproblem in the previous section. Basically when the matrix $[B]$ is an identity matrix, equation (11.3) becomes a standard eigenvalue problem dealt with in the previous section. So, the straightforward approach would be to pre-multiply equation (11.3) by $[B]^{-1}$ to bring it to the standard eigenvalue problem format of equation (11.4). However, this requires the inversion of a large matrix, which is not a computationally efficient approach. In addition, in structural dynamic applications the matrix $[B]$ would be the mass matrix $[M]$ of the dynamical system, and often may not be positive definite for it to be inverted. In order to overcome these difficulties a Lanczos recursion scheme that directly applies to the generalized eigenvalue problem is presented here.

Before presenting the recursion scheme we need to consider the transposed eigenproblem for the given problem. For convenience, the given eigenproblem is rewritten here.

$$[A]\{x_i\} = \lambda_i[B]\{x_i\} \tag{11.3}$$

The corresponding transposed eigenproblem is given by

$$[A]^{\mathrm{T}}\{z_i\} = \lambda_i[B]^{\mathrm{T}}\{z_i\} \tag{11.17}$$

It can be shown that the eigenvalues of the transposed eigenproblem is the same as that of the original problem [178] owing to the biorthonormal property of the right- and left-hand eigenvectors $\{x_i\}$ and $\{z_i\}$, respectively. The biorthonormality relationship is given by

$$\begin{aligned} \{z_i\}^{\mathrm{T}}[B]\{x_i\} &= 1 \\ \{z_j\}^{\mathrm{T}}[B]\{x_i\} &= 0 \quad (i \neq j) \end{aligned} \tag{11.18}$$

Now we are ready to set up the Lanczos two-sided recursion for the generalized non-symmetric eigenproblem.

Two arbitrarily chosen vectors $\{\hat{v}_1\}$ and $\{\hat{w}_1\}$ are binormalized with respect to the matrix $[B]$ to get the starting Lanczos vectors $\{v_1\}$ and $\{w_1\}$ yielding $\{w_1\}^{\mathrm{T}}[B]\{v_1\} = 1$. The binormalization of the randomly chosen vectors can be done by evaluating two scalar coefficients $\delta_0 = |\{\hat{w}_1\}^{\mathrm{T}}[B]\{\hat{v}_1\}|^{1/2}$ and $\beta_0 = \delta_0 \operatorname{sign}(\{\hat{w}_1\}^{\mathrm{T}}[B]\{\hat{v}_1\})$ and normalizing the vectors as follows:

$$\{v_1\} = \{\hat{v}_1\}/\delta_0 \qquad \text{and} \qquad \{w_1\} = \{\hat{w}_1\}/\beta_0 \tag{11.19}$$

The right and left Krylov sequence of vectors that will map the right- and left-hand eigenvectors of the problem, respectively, are given by

$$\{v_1, ([A]^{-1}[B])v_1, ([A]^{-1}[B])^2 v_1, \ldots, ([A]^{-1}[B])^{m-1} v_1\} \tag{11.20a}$$

$$\{w_1, ([A]^{-\mathrm{T}}[B]^{\mathrm{T}})w_1, ([A]^{-\mathrm{T}}[B]^{\mathrm{T}})^2 w_1, \ldots, ([A]^{-\mathrm{T}}[B]^{\mathrm{T}})^{m-1} w_1\} \tag{11.20b}$$

Two sets of vectors $[V] = [v_1, v_2, \ldots, v_m]$ and $[W] = [w_1, w_2, \ldots, w_m]$ are generated from the right and left Krylov sequence of vectors such that they satisfy the biorthonormal relationship

$$[W]^{\mathrm{T}}[B][V] = [I] \tag{11.21}$$

Starting from equation (11.19) the right and left Lanczos vectors $\{v_j\}$ and $\{w_j\}$ are generated employing the following steps.

For $j = 1, 2, \ldots, m \; (m < n)$

$$\{\overline{v}_{j+1}\} = [A]^{-1}[B]\{v_j\} - \alpha_j\{v_j\} - \beta_j\{v_{j-1}\} \tag{11.22}$$

$$\{\overline{w}_{j+1}\} = [A]^{-\mathrm{T}}[B]^{\mathrm{T}}\{w_j\} - \alpha_j\{w_j\} - \delta_j\{w_{j-1}\} \tag{11.23}$$

When $j = 1$, $\beta_1 v_0 = 0$ and $\delta_1 w_0 = 0$.

$$\alpha_j = \{w_j\}^{\mathrm{T}}[B][A]^{-1}[B]\{v_j\} \tag{11.24}$$

$$\delta_{j+1} = |\{\overline{w}_{j+1}\}^{\mathrm{T}}[B]\{\overline{v}_{j+1}\}|^{1/2} \tag{11.25}$$

$$\beta_{j+1} = \delta_{j+1} \operatorname{sign}(\{\overline{w}_{j+1}\}^{\mathrm{T}}[B]\{\overline{v}_{j+1}\}) \tag{11.26}$$

If at any step during the recursion $\delta_{j+1} = 0$, the recursion will have to be restarted with a new set of starting vectors $\{v_1\}$ and $\{w_1\}$.

$$\{v_{j+1}\} = \frac{\{\overline{v}_{j+1}\}}{\delta_{j+1}} \tag{11.27}$$

$$\{w_{j+1}\} = \frac{\{\overline{w}_{j+1}\}}{\beta_{j+1}} \tag{11.28}$$

The coefficients α_j, β_j and δ_j computed from the recursion presented in equations (11.19) and (11.22) through (11.28) are used to form the biorthogonally transformed matrix $[T]$ as shown in equation (11.16).

Since the eigenvalues are preserved under a biorthogonal transformation, the $[T]$ matrix will contain the eigenvalues of the original problem (equation 11.3). A proof of the transformation is shown here. In equation (11.3) let us set the right-hand eigenvector $\{x_i\}$ as

$$\{x_i\} = [V]\{y_i\} \tag{11.29}$$

Then pre-multiplying by $[W]^{\mathrm{T}}[B][A]^{-1}$ and after rearranging, equation (11.3) becomes

$$[W]^{\mathrm{T}}[B][A]^{-1}[B][V]\{y_i\} = (1/\lambda_i)[W]^{\mathrm{T}}[B][V]\{y_i\} \tag{11.30}$$

Using the biorthonormal property of the Lanczos vectors, namely $[W]^{\mathrm{T}}[B][V]=[I]$, we can rewrite equation (11.30) as

$$[T]\{y_i\} = \mu_i\{y_i\} \tag{11.31}$$

where

$$[T] = [W]^{\mathrm{T}}[B][A]^{-1}[B][V] \tag{11.32}$$

and

$$\mu_i = 1/\lambda_i \tag{11.33}$$

Equation (11.31) is the transformed eigenvalue problem whose eigenvalues and eigenvectors are related to the original problem through equations (11.33) and (11.29), respectively. In a computer implementation of the Lanczos method the tridiagonal $[T]$ matrix is directly formed using the scalar coefficients as given in equation (11.16). A proof showing how equation (11.32) yields the same $[T]$ matrix as in equation (11.16) can be found in the references [195] and [178] for symmetric and non-symmetric eigenvalue problems, respectively. It is evident from the two sided recursion presented that when the system matrices are symmetric, $[V]=[W]$ which leads to a single sided recursion for the symmetric eigenvalue problems.

When the eigenvalues and eigenvectors of the $[T]$ matrix are extracted, the eigenpairs of the original system can be recovered through equations (11.33) and (11.29), respectively. In general the $[T]$ matrix will be non-symmetric since $\beta_j = \pm\,\delta_j$, based on equation (11.26). A QR algorithm for non-symmetric matrices can be employed to extract the eigenvalues μ_i of the tridiagonal $[T]$ matrix quite efficiently. Complex arithmetic needs to be employed in the QR iterations since the eigenvalues of a non-symmetric matrix can be complex.

The size of the $[T]$ matrix will be equal to the number of pairs of Lanczos vectors m. We choose m much smaller than the size n of the original system. Therefore, the eigenvalues μ_i $(i = 1, 2, \ldots, m)$ of the $[T]$ matrix will be an approximation to m eigenvalues λ_i of the original system. As the size of $[T]$ matrix is increased, the μ_is will converge to yield λ_is that are close to the eigenvalues of the original problem. In practical implementations of the Lanczos scheme the number of Lanczos steps taken is limited by issues relating to the loss of biorthogonality of the vectors. Measures that need to be taken to deal with the loss of biorthogonality of the Lanczos vectors and the shift and search schemes are presented in the next section.

11.3.3. Implementation details of the Lanczos scheme

Inverse of [A] and $[A]^{\mathrm{T}}$

As pointed out in the previous section, the matrix $[A]$ in the recursion is not explicitly inverted in practical implementations. Also, the inverse of $[A]^{\mathrm{T}}$ can be recovered from the factorized $[A]$. While the inverse of $[A]^{\mathrm{T}}$ is not explicitly computed, the product $[A]^{-\mathrm{T}}[B]^{\mathrm{T}}\{w_j\}$ in equation (11.23) is obtained using the triangularized $[A]$ that is used to compute the product $[A]^{-1}\ [B]\ \{v_j\}$ in equation (11.22). In order to illustrate the procedure we consider the first term on the right-hand sides of equations (11.22) and (11.23).

$$\{\hat{v}\} = [A]^{-1}[B]\{v_j\} \tag{11.34a}$$

$$\{\hat{w}\} = [A]^{-\mathrm{T}}[B]^{\mathrm{T}}\{w_j\} \tag{11.34b}$$

Denoting the products $[B]\ \{v_j\}$ and $[B]^{\mathrm{T}}\{w_j\}$ by the vectors $\{p\}$ and $\{q\}$, respectively, the equations (11.34a) and (11.34b) are rewritten as

$$[A]\{\hat{v}\} = \{p\} \tag{11.35a}$$

$$[A]^{\mathrm{T}}\{\hat{w}\} = \{q\} \tag{11.35b}$$

The Gauss elimination process applied to the linear matrix equation (11.35a) involves reduction of the matrix $[A]$ into an upper triangular form and it can be symbolized as follows:

$$[L][D][S]\{\hat{v}\} = \{p\} \tag{11.36}$$

$\{L\}$ is a lower triangular matrix with unit diagonal elements. Its off-diagonal elements are composed of the multiplying factors used in the factorization of $[A]$ to get the upper triangular matrix $[S]$. The diagonal elements of $[S]$ are unity. $[D]$ is a diagonal matrix containing the pivot coefficients that arise in the Gauss elimination process. From equation (11.36) one can write the factored form of equation (11.35b) as follows:

$$[S]^{\mathrm{T}}[D][L]^{\mathrm{T}}\{\hat{w}\} = \{q\} \tag{11.37}$$

For a symmetric $[A]$ it is easy to see that $[L]=[S]^{\mathrm{T}}$ since $[A]^{\mathrm{T}}=[A]$. Owing to the non-symmetric $[A]$, we have to store both $[L]$ and $[S]$. Equations (11.36) and (11.37) are now rewritten as

$$[S]\{\hat{v}\} = [D]^{-1}[L]^{-1}\{p\} \tag{11.38a}$$

$$[L]^{\mathrm{T}}\{\hat{w}\} = [D]^{-1}[S]^{-T}\{q\} \tag{11.38b}$$

From equations (11.38), the solution vectors $\{\hat{v}\}$ and $\{\hat{w}\}$ can be computed through a simple back substitution process since $[S]$ and $[L]^{\mathrm{T}}$ are upper triangular matrices. Therefore, by using the triangular matrices $[L]$ and $[S]$ along with the pivot elements in $[D]$, the matrix multiplication involving $[A]^{-1}$ and $[A]^{-\mathrm{T}}$ in equations (11.22) and (11.23), respectively, can be computed in the two-sided recursion.

Reorthogonalization of Lanczos vectors

Loss of orthogonality of Lanczos vectors in finite precision computations is a deficiency of the recursion as presented in equations (11.22–11.28). Therefore, application of a reorthogonalization step to the Lanczos vectors $\{v_{j+1}\}$ and $\{w_{j+1}\}$ obtained in equations (11.27) and (11.28) is an essential part of the recursion to maintain a certain level of biorthogonality of the Lanczos vectors generated. Also, without any reorthogonalization an eigenvector that converges at the jth Lanczos step will gradually grow into the Lanczos vectors computed in the subsequent steps of the recursion, causing an already converged eigenvalue to be detected again. Different schemes of reorthogonalizing the current Lanczos vector against the already computed ones are presented in reference [195]. Either the full or the selective reorthogonalization scheme can be used to maintain orthogonality between all the computed Lanczos vectors. In the literature the selective reorthogonalization scheme has been shown to be economical in computations when the number of Lanczos step taken is large, say more than 50. However, in this presentation of the method a full reorthogonalization scheme is shown which is tailored for the two-sided recursion where fewer than 50 or so Lanczos steps are taken. To ensure biorthogonality of the pair of Lanczos vectors computed at the $(j+1)$th step, the Gram–Schmidt coefficients are computed using the following expressions.

For $i = 1, 2, \ldots, j$

$$\theta_i = \{w_i\}^{\mathrm{T}}[B]\{v_{j+1}\} \tag{11.39a}$$

$$\phi_i = \{v_i\}^{\mathrm{T}}[B]^{\mathrm{T}}\{w_{j+1}\} \tag{11.39b}$$

If any of the coefficients is, say θ_k and/or ϕ_k, is greater than a predetermined small number ε_0, then $\{v_{j+1}\}$ and $\{w_{j+1}\}$ are modified using the following Gram–Schmidt biorthogonalization steps:

$$\{v_{j+1}\} \to \{v_{j+1}\} - \theta_k\{v_k\} \tag{11.40a}$$

$$\{w_{j+1}\} \to \{w_{j+1}\} - \phi_k\{w_k\} \tag{11.40b}$$

The value of the biorthogonality tolerance ε_0 is chosen based on the machine round-off error. While the level of biorthogonality of Lanczos vectors is dependent upon the tolerance ε_0, it also affects the ability of the recursion to capture the repeated eigenvalues if any are present in a problem. Therefore the choice of the value of ε_0 should be done with care. A value of 10^{-8} has been found to work well based on the above two requirements on the tolerance value. The number of steps it takes to capture a repeated mode is problem dependent. In the numerical studies conducted with acoustic fluid–structure eigenvalue problems, a repeated mode usually converged in about five to ten Lanczos steps after the first has converged. However, eigenvalues that are complex always converged in complex conjugate pairs. In reference [196] Kim and Craig use a block Lanczos algorithm in order to capture the repeated eigenvalues. By choosing a block size equal to or larger than the multiplicity of the eigenvalues, the repeated modes are found more reliably.

Eigenvalue search strategy using shift logic

In the two-sided recursion applied to the standard eigenvalue problem, equation (11.4), the Lanczos scheme has the property of quickly converging to the largest eigenvalues of the matrix $[A]$. Owing to the transformation that takes place in the generalized eigenproblem recursion, which involves $[A]^{-1}$, the lowest eigenvalues converge first as can be seen from the relationship $\mu_i = 1/\lambda_i$ in equation (11.33).

The convergence of eigenvalues in a specified part of the eigenvalue spectrum can be accelerated by applying an appropriate shift to the given eigenproblem. A spectral transformation of equation (11.3) is introduced by setting

$$\lambda_i = \nu_i + \sigma \tag{11.41}$$

The spectral transformed eigenvalue problem is then given by

$$([A] - \sigma[M])\{x_i\} = \nu_i[B]\{x_i\} \tag{11.42}$$

In the non-symmetric eigenvalue problem with complex eigenvalues, this transformation accelerates convergence of eigenvalues whose magnitude is close to that of the shift σ. Through numerical experimentation it has been observed that the shift needs to be only a real number even when complex eigenvalues are expected. The value of σ is chosen close to the magnitude of the complex eigenvalues searched.

The following approach is used to search for the eigenvalues in a specified part of the eigenvalue spectrum. A shift σ equal to the lowest eigenvalue magnitude sought is applied to the problem. If the number of eigenvalues requested is r, $m = r + s$ recursion steps are performed. The value of s is chosen such that r eigenvalues with magnitude beyond the shift would converge without missing any modes. After each Lanczos recursion step, the eigenvalues are computed and the converged ones accumulated. At the end of m steps, the first r eigenvalues with magnitude higher than the shift, arranged in ascending order of their magnitude, are presented as converged eigenvalues. After some numerical experimentation a value of $s = 16$ was chosen. However, to ensure that no eigenvalues are missed a Sturm sequence check, as presented in reference [186], should be employed combined with the strategy outlined. For the non-symmetric eigenproblems, however, a Sturm sequence check procedure is unavailable to the knowledge of the authors of this book.

11.4. Example problems

We present two example problems as a demonstration of the working of the two-sided Lanczos recursion presented. The first one is a generalized eigenproblem of 12×12 matrices $[A]$ and $[B]$ with known eigenvalues that are complex. The second example is an acoustic fluid–structure problem where we are interested in finding the eigenvalues of a fluid filled cylindrical shell for which the theoretical resonant frequencies are available for comparison.

Example 11.1: Eigenvalues of a non-symmetric matrix

In this example the $[A]$ and $[B]$ matrices of the generalized eigenproblem were generated from two 12×12 matrices where one is a block-diagonal matrix of block size 2 and the second one is an identity matrix. The 2×2 blocks of the block-diagonal matrix was filled with the real and imaginary parts of the complex conjugate eigenvalue pair chosen a priori. The diagonals of a 2×2 block are filled with the real part of a complex eigenvalue and the off-diagonals with the imaginary part forming a skew symmetric matrix. The matrices $[A]$ and $[B]$ were obtained from the first and the second matrices, respectively, by pre-multiplying them with an arbitrarily chosen matrix filled with random numbers. So, the eigenvalues of the generalized eigenvalue problem is known beforehand for comparing with the computed eigenvalues employing the Lanczos

Table 11.1. Matrices $[A]$ and $[B]$ used in Example 11.1.

$[A]$ Matrix

−41	176	393	42	14	616	−700	−1270	5333	2376	8320	−1400
34	418	477	225	14	378	1685	−315	−455	5199	−11960	5200
−4	−924	−396	354	−315	574	590	660	3191	−54	6760	−6000
−11	−330	180	−555	−280	182	90	1550	−3712	−5844	3120	30000
19	−836	273	21	112	−238	−70	400	−968	−826	−1560	76000
3	44	75	−159	217	518	−370	−1310	−2273	5722	8320	−98000
11	1056	−648	−399	56	462	−375	625	5493	−164	10140	8000
34	286	−300	−537	−301	−644	−815	−285	2125	4724	−8840	−12000
−49	−748	−576	93	−21	336	900	540	6256	−2156	9620	96000
−3	594	−39	48	−273	154	−280	920	1436	1493	10660	90000
−17	110	261	−690	91	−140	−210	1200	−2629	3277	11180	−24000
29	−44	−69	642	−203	350	−555	−1285	4944	2479	5720	−36000

$[B]$ Matrix

−410	16	90	33	2	88	125	−215	−16	43	64	7
340	38	99	81	2	54	195	220	−41	−1	−92	−26
−40	−84	−114	60	−45	82	−45	145	2	25	52	3
−110	−30	75	−120	−40	26	−220	150	44	−32	24	−15
190	−76	63	21	16	−34	−65	25	6	−8	−12	−38
30	4	27	−33	31	74	160	−170	−46	−15	64	49
110	96	−129	−132	8	66	−125	0	4	43	78	−4
340	26	−39	−144	43	−92	−30	−145	−36	19	−68	6
−490	−68	−141	−12	−3	48	0	180	20	48	74	−48
−30	54	−12	9	−39	22	−160	40	−11	12	82	−45
−170	10	102	−147	13	−20	−195	75	−27	−19	86	12
290	−4	−54	147	−29	50	140	−195	−17	40	44	18

two-sided recursion presented in the previous section. The matrices $[A]$ and $[B]$ are shown in Table 11.1.

This generalized eigenproblem has both real and complex eigenvalues. The computed and known eigenvalues are presented in Table 11.2. Since the problem considered here is quite small, the Lanczos recursion was performed to the full size of the system ($m = n = 12$), transforming $[A]$ and $[B]$ into a subspace size of 12 yielding a 12×12 $[T]$ matrix. It is seen from the table of results that the eigenvalues of the non-symmetric generalized eigenproblem are quite accurately computed by the Lanczos method presented. In general for large eigenvalue problems the number of Lanczos steps taken can be much smaller than the size of the matrices $[A]$ and $[B]$ in order to extract the first few eigenvalues of the problem. The next example is designed to illustrate this point.

Example 11.2: Fluid–structure eigenvalue problem

This example is an illustration of a practical application of the non-symmetric generalized eigenvalue solution method presented in the previous section. The problem considered is an acoustic fluid–structure interaction problem. The eigenvalues of this problem will all be real values even though the stiffness and mass matrices of the problem, $[K]$ and $[M]$, respectively, are non-symmetric. A cylindrical shell of circular cross section filled with fluid is analyzed employing the Finite Element method (FEM). The length of the shell is assumed to be infinite. Shown in Figure 11.1 is the cross section

Table 11.2. Eigenvalues of the input matrix problem in example 11.1.

No.	Known eigenvalues	Computed eigenvalues	
1	0.1	0.100000	+0.000000J
2	4 − 1J	4.000000	−1.000000J
3	4 + 1J	4.000000	+1.000000J
4	3 − 5J	3.000000	−5.000000J
5	3 + 5J	3.000000	+5.000000J
6	7	7.000000	−0.000000J
7	7	7.000000	+0.000000J
8	11	11.000000	+0.000000J
9	8 − 127J	8.000000	−127.000000J
10	8 + 127J	8.000000	+127.000000J
11	130	129.99989	+0.000000J
12	−2000	−2000.022577	+0.000000J

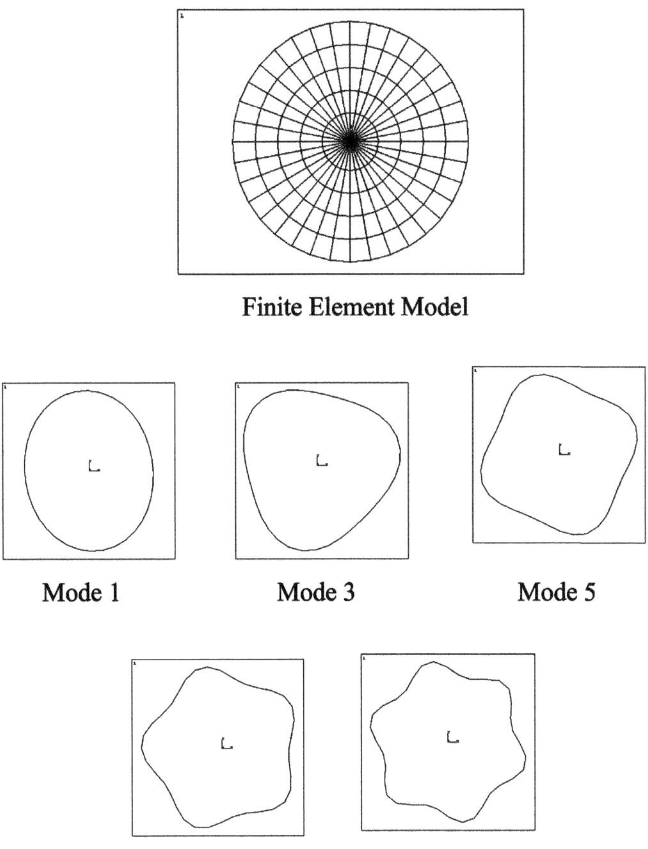

Figure 11.1. Fluid-filled Shell model and mode shapes (Shell: wall thickness $t = 0.102$ m, Young's modulus $E = 2.068e11$ N/m^2, Poisson's ratio $\nu = 0$, density $\rho_s = 2403$ kg/m^3; fluid: bulk modulus $K = 2.25e9$ N/m^2, density $\rho = 1000$ kg/m^3; modes 2, 4, 6, 8, 10 have the same shape but are oriented at 90 degrees to modes 1, 3, 5, 7, 9 respectively).

of the finite element discretized shell and the fluid inside. The material properties of the steel shell and the water inside are shown in the figure.

In Section 10.2.1 the same problem was considered to show the application of the boundary element eigenvalue problem formulation developed in Chapters 7 through 10. In this current chapter, however, it is used to demonstrate application of the non-symmetric eigenvalue solver.

Employing the FEM, the dynamic matrix equation for free vibration analysis of an acoustic fluid–structure interaction problem is shown below in equation (11.44). The shell element and the acoustic fluid element available in the ANSYS® general-purpose finite element program [160] are used. The shell element is a four-node element allowing cubic variation of the radial displacement along circumference and constant displacement through the thickness. The acoustic element is a eight-node three-dimensional (3-D) brick with tri-linear pressure shape functions representing the acoustic wave equation (5.1). Even though 3-D elements are used, the analysis is carried out in 2-D to treat the planar cylindrical ring problem. The free vibration response equation of the coupled problem is rewritten from Chapter 10 here. The structural dynamic equation (10.4) is combined with the following fluid finite element matrix equation from Zienkiewicz and Newton [186]

$$-\omega^2[M_{\mathrm{f}}]\{P\} + [K_{\mathrm{f}}]\{P\} = \rho\omega^2[R]\{U\} \tag{11.43}$$

to form the fluid–structure matrix equation given by

$$\begin{bmatrix} K_{\mathrm{s}} & -R \\ 0 & -K_{\mathrm{f}} \end{bmatrix} \begin{Bmatrix} U \\ P \end{Bmatrix} = \omega_i^2 \begin{bmatrix} M_{\mathrm{s}} & 0 \\ \rho R^{\mathrm{T}} & M_{\mathrm{f}} \end{bmatrix} \begin{Bmatrix} U \\ P \end{Bmatrix} \tag{11.44}$$

$\{U\}$ and $\{P\}$ contain the unknown nodal displacements and pressures respectively. $[K_{\mathrm{s}}]$ and $[M_{\mathrm{s}}]$ are the stiffness and mass matrices of the shell. $\rho[R^{\mathrm{T}}]$ is the matrix generated at the fluid–shell interface using the interface condition given by equation (10.1), namely the fluid pressure gradient at the interface balances the acceleration force exerted by the structure. ρ is the density of the fluid. $[K_{\mathrm{f}}]$ and $[M_{\mathrm{f}}]$ are the fluid "stiffness" and "mass" matrices, respectively. This is a non-symmetric matrix generalized eigenproblem even though the sub-matrices are symmetric except for the coupling submatrix $[R]$. The two-sided Lanczos recursion was applied to this eigenvalue problem to extract the eigenvalues $\lambda_i = \omega_i^2$.

The first ten non-zero frequencies, $f_i = \omega_i/2\pi$, extracted are presented in Table 11.3. Due to the geometric symmetry of the problem the eigenvalues appear in pairs which correspond to the two eigenmodes oriented about two orthogonal diameters of the ring. The five mode shapes computed are also shown in Figure 11.1. The first two pairs of eigenvalues are in good agreement with the theoretical solution presented by Yu [188]. At higher modes the computed frequencies slightly differ from the theoretical frequencies which may be attributed to the Finite Element discretization approximation of the continuum.

11.5. Summary statements on the non-symmetric Lanczos eigensolver

The Lanczos two-sided algorithm presented in Sections 11.2 through 11.4 is given in its most general form. Therefore it can be applied to extract both the real and

Table 11.3. Eigenfrequencies of the fluid-filled shell.

	Frequency (Hz)	
Mode	Theoretical	Computed
1	25.038	25.025378
2	25.038	25.025378
3	77.924	78.380262
4	77.924	78.380263
5	160.42	162.55957
6	160.42	162.55959
7	272.99	278.84044
8	272.99	278.84045
9	415.17	427.61895
10	415.17	427.61924

complex eigenvalues of a non-symmetric system. For the Boundary Element algebraic eigenvalue problem this method is ideally suited owing to its generality of the formulation to handle non-symmetric matrices with the possibility of complex eigenvalues that may arise due to discretization approximation of the continuum.

For short Lanczos runs where only a few eigenvalues of a large problem are sought, the computation speed was found to be quite fast compared to other subspace procedures. When a large number of Lanczos steps are taken in order to compute several eigenvalues, say more than 100 modes, a selective reorthogonalization procedure will help improve the computational performance of the method.

11.6. Damped system eigenvalue problem solution

In Section 10.3 we described the inclusion of sound absorption in acoustic boundary element eigenvalue formulations. Also, in structural dynamic applications involving damped systems a damping matrix would arise in the eigenvalue formulations leading to a quadratic eigenvalue problem presented in equation (10.22). This quadratic eigenvalue problem equation is rewritten here.

$$[A]\{x_i\} + \lambda_i[C]\{x_i\} + \lambda_i^2[B]\{x_i\} = \{0\} \qquad (11.45)$$

In dynamic applications, $[A]$ and $[B]$ represent the stiffness and mass matrices, respectively. The matrix $[C]$ is the damping matrix, which accounts for the dissipation of energy. When $[C]$ is zero, equation (11.45) reduces to the generalized eigenvalue problem for which the Lanczos recursion was presented in Section 11.3. In the damped system boundary element eigenvalue problem these matrices are non-symmetric and in practical situations their size n can be large. As we showed in the previous sections, the Lanczos subspace approach would be best suited for such systems where only a few eigenvalues at one end of the spectrum, usually the lowest ones, are of interest.

One of the straightforward approaches is to employ the method of matrix augmentation which leads to a linear eigenvalue problem of the form in equation (11.3). Then the Lanczos two-sided recursion presented in Section 11.3 can be employed to solve for the eigenvalues and vectors of the problem. Defining a new vector $\{\tilde{x}_i\} = \lambda_i\{x_i\}$, we can write a null equation.

$$[B]\{\tilde{x}_i\} - [B]\lambda_i\{x_i\} = \{0\} \tag{11.46}$$

Combining it with equation (11.45), the augmented linear eigenproblem given below can be arrived at.

$$\begin{bmatrix} A & 0 \\ 0 & -B \end{bmatrix} \begin{Bmatrix} x_i \\ \tilde{x}_i \end{Bmatrix} = \lambda_i \begin{bmatrix} -C & -B \\ -B & 0 \end{bmatrix} \begin{Bmatrix} x_i \\ \tilde{x}_i \end{Bmatrix} \tag{11.47}$$

Solution of this linear eigenproblem involves matrices that are doubled in size to $2n$ and requires double the computational effort. Therefore, in the sections that follow the two-sided Lanczos recursion that applies directly to non-symmetric quadratic eigenvalue problem given by equation (11.45) is developed.

11.7. Lanczos two-sided recursion for the quadratic eigenvalue problem

Pursuing along the same lines of Section 11.3 the Lanczos two-sided recursion scheme to compute the eigenvalues of the quadratic eigenvalue problem is shown in this section. In the development of the algorithm we will make use of the biorthonormal transformation of the quadratic eigenproblem that leads to the tridiagonal subspace matrix [eqn. (11.16)]. As we saw in Section 11.3, the two-sided recursion is developed by considering the original eigenproblem and its transpose in order to generate two sets of vectors namely the right and left hand Lanczos vectors. The transposed quadratic eigenvalue problem is written below.

$$[A]^{\mathrm{T}}\{z_i\} + \lambda_i[C]^{\mathrm{T}}\{z_i\} + \lambda_i^2[B]^{\mathrm{T}}\{z_i\} = \{0\} \tag{11.48}$$

Basically, the zeroes of the determinant equation

$$|[A]^{\mathrm{T}} + \lambda_i[C]^{\mathrm{T}} + \lambda_i^2[B]^{\mathrm{T}}| = 0 = |([A] + \lambda_i[C] + \lambda_i^2[B])^{\mathrm{T}}| \tag{11.49}$$

are the eigenvalues λ_i, of equation (11.48). Since the determinant of a matrix is the same as that of its transpose, it is clear that the eigenvalues of the original and the transposed problems, given by equations (11.45) and (11.48), respectively are the same. The associated eigenvectors $\{x_i\}$ and $\{z_i\}$ are the right- and left-hand eigenvectors. Therefore, the Lanczos recursion is derived considering the original eigenproblem in equation (11.45) and its transposed problem in equation (11.48) in order to biorthogonally transform the n order quadratic problem into a standard eigenvalue problem (equation 11.31) of order m, $m < n$, in the subspace of the Lanczos vectors.

11.7.1. Biorthogonality relationship for the quadratic eigenvalue problem

The biorthogonality relationship of the right and left hand eigenvectors can be derived by considering two distinct eigenvalues λ_i and λ_j, $i \neq j$, in equations (11.45) and (11.48), respectively. Corresponding to the eigenvalue λ_j, equation (11.48) is written as

$$[A]^{\mathrm{T}}\{z_j\} + \lambda_j[C]^{\mathrm{T}}\{z_j\} + \lambda_j^2[B]^{\mathrm{T}}\{z_j\} = \{0\} \tag{11.50}$$

Pre-multiplying equation (11.45) by $\{z_j\}^{\mathrm{T}}$ and equation (11.50) by $\{x_i\}^{\mathrm{T}}$ and transposing it, the difference between the two resulting equations is written down below.

$$(\lambda_i - \lambda_j)\{z_j\}^{\mathrm{T}}[C]\{x_i\} + (\lambda_i^2 - \lambda_j^2)\{z_j\}^{\mathrm{T}}[B]\{x_i\} = 0 \tag{11.51}$$

Factoring out $(\lambda_i - \lambda_j)$ in equation (11.51), and noting that $\lambda_i \neq \lambda_j$, the biorthogonality relationship is found to be

$$\{z_j\}^{\mathrm{T}}[C]\{x_i\} + (\lambda_i + \lambda_j)\{z_j\}^{\mathrm{T}}[B]\{x_i\} = 0 \tag{11.52}$$

Here, we introduce two additional eigenvectors as we did to arrive at the linearized eigenproblem in equation (11.47).

$$\{\tilde{x}_i\} = \lambda_i\{x_i\} \tag{11.53a}$$

$$\{\tilde{z}_j\} = \lambda_j\{z_j\} \tag{11.53b}$$

Equation (11.52) can now be written as

$$\{z_j\}^{\mathrm{T}}[C]\{x_i\} + \{\tilde{z}_j\}^{\mathrm{T}}[B]\{x_i\} + \{z_j\}^{\mathrm{T}}[B]\{\tilde{x}_i\} = 0 \tag{11.54}$$

The generalized biorthogonality condition for the quadratic eigenproblem, therefore, is given by equation (11.54). When $j = i$, if the eigenvectors are normalized such that

$$\{z_i\}^{\mathrm{T}}[C]\{x_i\} + \{\tilde{z}_i\}^{\mathrm{T}}[B]\{x_i\} + \{z_i\}^{\mathrm{T}}[B]\{\tilde{x}_i\} = 1 \tag{11.55}$$

The generalized biorthonormal condition of the quadratic eigenproblem is written down as follows:

$$[Z]^{\mathrm{T}}[C][X] + [\tilde{Z}]^{\mathrm{T}}[B][X] + [Z]^{\mathrm{T}}[B][\tilde{X}] = [I] \tag{11.56}$$

Normalization given by equation (11.55) may not always be possible due to the possibility that the sum of the three scalar products may yield a null value. In equation (11.56), $[X] = [x_1, x_2, \ldots, x_n]$ and $[Z] = [z_1, z_2, \ldots, z_n]$ are the right- and left-hand eigenvector matrices. For a given eigenvalue λ_i, the eigenvectors defined in equation (11.53) are obtainable from $\{x_i\}$ and $\{z_i\}$. Therefore, $[\tilde{X}] = [\tilde{x}_1, \tilde{x}_2, \ldots, \tilde{x}_n]$ and $[\tilde{Z}] = [\tilde{z}_1, \tilde{z}_2, \ldots, \tilde{z}_n]$ are called the right- and left-hand dependent eigenvector matrices of the quadratic eigenproblem. From the biorthonormal relationship established here it is evident that the Lanczos recursion for the quadratic eigenproblem will need to generate two independent and two dependent sets of right- and left-hand vectors in order to biorthogonally transform the eigenproblem.

11.7.2. Lanczos recursion

With the definition of the biorthonormality expressions given by equation (11.56) we are ready to develop the Lanczos recursion scheme for the non-symmetric quadratic eigenvalue problem posed in equations (11.45) and (11.48) earlier in this section. The Lanczos recursion for this case is developed in detail with complete proofs and derivations by Rajakumar [179]. Here we will present the algorithm from the point of view of practical implementation of the recursion, without delving into to detailed derivations.

Again, along the lines of Section 11.3.2, we start by seeking the Krylov sequence of vectors that will generate the Lanczos vectors. The eigenproblem equations (11.45) and (11.48) can be rewritten using the definition of the dependent eigenvectors in equation (11.53) as follows:

$$[A]\{x_i\} = -\lambda_i([C]\{x_i\} + [B]\{\tilde{x}_i\}) \tag{11.57a}$$

$$[A]^{\mathrm{T}}\{z_i\} = -\lambda_i([C]^{\mathrm{T}}\{z_i\} + [B]^{\mathrm{T}}\{\tilde{z}_i\}) \tag{11.57b}$$

Now, two sets of arbitrarily chosen vectors ($\{\hat{v}_1\}$, $\{\hat{w}_1\}$) and ($\{\hat{r}_1\}$, $\{\hat{s}_1\}$) are used to start the two pairs of Krylov sequence of vectors that apply to equations (11.45) and (11.48). From equations (11.57), the right- and left-hand sequence of vectors that will map the independent eigenvectors $\{x_i\}$ and $\{z_i\}$, respectively, are

$$\{\{\hat{v}_1\}, \{\hat{v}_2\} = -[A]^{-1}([C]\{\hat{v}_1\} + [B]\{\hat{r}_1\}), \{\hat{v}_3\} = -[A]^{-1}([C]\{\hat{v}_2\} + [B]\{\hat{r}_2\}), \ldots\} \tag{11.58a}$$

$$\begin{aligned}\{\{\hat{w}_1\}, \{\hat{w}_2\} &= -[A]^{-\mathrm{T}}([C]^{\mathrm{T}}\{\hat{w}_1\} + [B]^{\mathrm{T}}\{\hat{s}_1\}), \{\hat{w}_3\} \\ &= -[A]^{-\mathrm{T}}([C]^{\mathrm{T}}\{\hat{w}_2\} + [B]^{\mathrm{T}}\{\hat{s}_2\}), \ldots\}\end{aligned} \tag{11.58b}$$

From equations (11.53), the right- and left-hand sequences that will map the dependent eigenvectors $\{\tilde{x}_i\}$ and $\{\tilde{z}_j\}$, respectively, can be written as

$$\{\{\hat{r}_1\}, \{\hat{r}_2\} = \{\hat{v}_1\}, \{\hat{r}_3\} = \{\hat{v}_2\}, \ldots\} \tag{11.59a}$$

$$\{\{\hat{s}_1\}, \{\hat{s}_2\} = \{\hat{w}_1\}, \{\hat{s}_3\} = \{\hat{w}_2\}, \ldots\} \tag{11.59b}$$

Inspecting equations (11.58) and (11.59), we can see that the primary Krylov sequences are coupled to the secondary sequences. These above Krylov sequences can also be written down by looking at the linearized augmented form of the quadratic eigenproblem given in equation (11.47). Now we are ready to proceed to present the Lanczos recursion for the problem.

Two sets of vectors $\{\hat{v}_1\}$, $\{\hat{w}_1\}$ and $\{\hat{r}_1\}$, $\{\hat{s}_1\}$ are arbitrarily chosen to start the recursion. Since they are arbitrary, we set $\{\hat{v}_1\} = \{\hat{w}_1\}$ and $\{\hat{r}_1\} = \{\hat{s}_1\}$. The first two sets of Lanczos vectors $\{v_1\}$, $\{w_1\}$ and $\{r_1\}$, $\{s_1\}$ are obtained from the starting vectors through normalization such that they are binormal yielding $\{w_1\}^{\mathrm{T}}[C]\{v_1\} + \{s_1\}^{\mathrm{T}}[B]\{v_1\} + \{w_1\}^{\mathrm{T}}[B]\{r_1\} = 1$. This normalization is done by dividing $\{\hat{v}_1\}$ and $\{\hat{r}_1\}$ by δ_1 and by dividing $\{\hat{w}_1\}$ and $\{\hat{s}_1\}$ by β_1 where the scalar coefficients are $\delta_1 = |\Delta_1|^{1/2}$, $\beta_1 = \delta_1$ sign(Δ_1), $\Delta_1 = \{\hat{w}_1\}^{\mathrm{T}}[C]\{\hat{v}_1\} + \{\hat{s}_1\}^{\mathrm{T}}[B]\{\hat{v}_1\} + \{\hat{w}_1\}^{\mathrm{T}}[B]\{\hat{r}_1\}$. Then the recursion steps are for $j = 1, 2, \ldots, m\,(m \le n)$,

$$\{\bar{v}_{j+1}\} = -[A]^{-1}([C]\{v_j\} + [B]\{r_j\}) - \alpha_j\{v_j\} - \beta_j\{v_{j-1}\} \tag{11.60a}$$

$$\{\overline{w}_{j+1}\} = -[A]^{-\mathrm{T}}([C]^{\mathrm{T}}\{w_j\} + [B]^{\mathrm{T}}\{s_j\}) - \alpha_j\{w_j\} - \delta_j\{w_{j-1}\} \tag{11.60b}$$

$$\{\bar{r}_{j+1}\} = \{v_j\} - \alpha_j\{r_j\} - \beta_j\{r_{j-1}\} \tag{11.61a}$$

$$\{\bar{s}_{j+1}\} = \{w_j\} - \alpha_j\{s_j\} - \delta_j\{s_{j-1}\} \tag{11.61b}$$

(when $j = 1$, $\beta_1\{v_0\} = \beta_1\{r_0\} = \delta_1\{w_0\} = \delta_1\{s_0\} = 0$)

$$\alpha_j = -(\{w_j\}^{\mathrm{T}}[C] + \{s_j\}^{\mathrm{T}}[B])[A]^{-1}([C]\{v_j\} + [B]\{r_j\}) + \{w_j\}^{\mathrm{T}}[B]\{v_j\} \tag{11.62}$$

$$\Delta_{j+1} = \{\overline{w}_{j+1}\}^{\mathrm{T}}[C]\{\bar{v}_{j+1}\} + \{\bar{s}_{j+1}\}^{\mathrm{T}}[B]\{\bar{v}_{j+1}\} + \{\overline{w}_{j+1}\}^{\mathrm{T}}[B]\{\bar{r}_{j+1}\} \tag{11.63}$$

$$\delta_{j+1} = |\Delta_{j+1}|^{1/2} \tag{11.64}$$

$$\beta_{j+1} = \delta_{j+1}\,\mathrm{sign}(\Delta_{j+1}) \tag{11.65}$$

$$\{v_{j+1}\} = \frac{\bar{v}_{j+1}}{\delta_{j+1}}, \quad \{r_{j+1}\} = \frac{\bar{r}_{j+1}}{\delta_{j+1}} \tag{11.66a}$$

$$\{w_{j+1}\} = \frac{\overline{w}_{j+1}}{\beta_{j+1}}, \quad \{s_{j+1}\} = \frac{\bar{s}_{j+1}}{\beta_{j+1}} \tag{11.66b}$$

The recursion scheme presented above in equations (11.60) through (11.66) generates an extra set of right- and left-hand Lanczos vectors as compared to the linear non-symmetric generalized eigenvalue problem shown in Section 11.3. The Lanczos vectors generated in the recursion scheme presented above satisfy the generalized biorthonormality condition given by

$$[W]^{\mathrm{T}}[C][V]+[S]^{\mathrm{T}}[B][V]+[W]^{\mathrm{T}}[B][R]=[I] \tag{11.67}$$

The coefficients generated in the recursion α_j, β_j, and δ_j are used to form the tridiagonal matrix $[T]$ as shown in equation (11.16) whose eigenvalues approximate the eigenvalues of the quadratic eigenproblem in equation (11.45). So, the quadratic eigenproblem reduces to solving the following standard eigenproblem

$$[T]\{y_i\}=\mu_i\{y_i\} \tag{11.68}$$

in the Lanczos vectors subspace of size m. The eigenvalues and eigenvectors of $[T]$ matrix and those of the original quadratic eigenproblem are related by $\mu_i=1/\lambda_i$ and $\{y_i\}=[V]\{x_i\}$. As the subspace size m is progressively increased, the eigenvalues of equation (11.68) converge to yield the eigenvalues λ_i of the quadratic eigenvalue problem. Theoretically, when $m=2n$ all of the eigenvalues of the quadratic eigenproblem will be extracted. However, often in practical applications involving large matrices only the first few eigenvalues are of interest which will converge in a few Lanczos recursion steps $m \ll 2n$. Spectral transformation of the eigenproblem to improve convergence of the eigenvalue in a specified part of the eigenvalue spectrum is presented in the next section dealing with practical implementation details.

11.7.3. Implementation details of quadratic Lanczos scheme

Proof of biorthogonal transformation

First we present the biorthogonal transformation proof that leads to the tridiagonal matrix $[T]$ for the quadratic Lanczos recursion scheme. The right hand eigenvectors are projected onto the subspace spanned by the Lanczos vectors as follows:

$$\{x_i\}=[V]\{y_i\}, \quad \{\tilde{x}_i\}=[R]\{y_i\} \tag{11.69}$$

Substituting equation (11.69) into (11.53a) and (11.57a), we multiply (11.57a) by $([W]^{\mathrm{T}}[C]+[S]^{\mathrm{T}}[B])[A]^{-1}$ and (11.53a) by $[W]^{\mathrm{T}}[B]$, and add the two equations to get

$$\begin{aligned}&[[W]^{\mathrm{T}}[C][V]+[S]^{\mathrm{T}}[B][V]+[W]^{\mathrm{T}}[B][R]+\lambda_i([W]^{\mathrm{T}}[C]\\&\quad+[S]^{\mathrm{T}}[B])[A]^{-1}([C][V]+[B][R])-\lambda_i[W]^{\mathrm{T}}[B][V]]=[0]\end{aligned} \tag{11.70}$$

Since the Lanczos vectors are biorthonormal, satisfying equation (11.67), we obtain the tridiagonal subspace eigenproblem

$$[T]\{y_i\}=(1/\lambda_i)\{y_i\} \tag{11.71}$$

where $[T]$ is the subspace matrix given by

$$[T]=-([W]^{\mathrm{T}}[C]+[S]^{\mathrm{T}}[B])[A]^{-1}([C][V]+[B][R])+[W]^{\mathrm{T}}[B][V] \tag{11.72}$$

Equation (11.71), which is the same as shown in (11.68), is the biorthogonally transformed eigenproblem whose eigenvectors relate to the original eigenproblem through (11.69). In the Lanczos eigenvalue solution process the $[T]$ matrix is directly formed using the coefficients generated in the recursion steps in equations (11.62) through

(11.65). The proof showing that equation (11.72), in fact, results in a tridiagonal matrix (11.16) is given in reference [163].

Spectral transformation

In computer implementation of the algorithm, to accelerate convergence of the eigenvalues in a specified eigenvalue magnitude spectrum, a spectral transformation needs to be incorporated. A shift of the eigenvalues is introduced as shown here.

$$\lambda_i = \nu_i + \sigma \tag{11.73}$$

Substituting equation (11.73) into (11.45) leads to the spectrally transformed eigenproblem given by

$$[A_\sigma]\{x_i\} + \nu_i[C_\sigma]\{x_i\} + \nu_i^2[B]\{x_i\} = \{0\} \tag{11.74}$$

where $[A_\sigma] = ([A] + \sigma[C] + \sigma^2[B])$ and $[C_\sigma] = ([C] + 2\sigma[B])$. The Lanczos algorithm has the property of converging first to the largest magnitude eigenvalues in the spectrum. Therefore, in view of the eigenvalue transformation equation, $\mu_i = 1/\lambda_i$, the lowest magnitude eigenvalues of the quadratic eigenvalue problem will converge first. The use of a real valued shift σ accelerates convergence of the eigenvalues that are close to its magnitude. Also, the shifting is needed to extract the rigid body modes in structural dynamic applications.

Re-biorthogonalization of Lanczos vectors

Equations (11.60) and (11.61) are the biorthogonalization steps of the recursion, where the set of $(j+1)$st vectors are biorthogonalized against the jth and $(j-1)$st vectors. These two steps are responsible for maintaining biorthogonality of Lanczos vectors such that the biorthogonality equation (11.67) is satisfied. However, as pointed out in Section 11.3.3 for the generalized non-symmetric eigenproblem, the Lanczos vectors gradually loose biorthogonality as the recursions proceed in computer implementations. This loss of biorthogonality occurs due to the limitation of the finite precision calculations that the computers employ. In exact arithmetic the recursion scheme as presented produces biorthonormal vectors. In other words, if the computers were to handle arithmetic calculations in infinite precision, i.e., in exact arithmetic, the Lanczos vectors as generated will be biorthonormal. Due to this practical limitation, a re-biorthogonal scheme to restore biorthogonality of the vectors is necessary.

The $(j+1)$st vectors computed in equations (11.66) are checked for their of level of biorthogonality with respect to all of the previous vectors that have been generated. For $i = 1, 2, \ldots, j$, the biorthogonality coefficients are calculated as follows:

$$\theta_i = \{w_i\}^{\mathrm{T}}[C]\{v_{j+1}\} + \{s_i\}^{\mathrm{T}}[B]\{v_{j+1}\} + \{w_i\}^{\mathrm{T}}[B]\{r_{j+1}\} \tag{11.75a}$$

$$\phi_i = \{w_{j+1}\}^{\mathrm{T}}[C]\{v_i\} + \{w_{j+1}\}^{\mathrm{T}}[B]\{r_i\} + \{s_{j+1}\}^{\mathrm{T}}[B]\{v_i\} \tag{11.75b}$$

If the magnitude of any of the coefficients, say, the kth coefficient, exceeds a predetermined small value ε_0, then the vectors are biorthogonalized with respect to the kth set of Lanczos vectors.

$$\{v_{j+1}\} \rightarrow \{v_{j+1}\} - \theta_k\{v_k\}, \quad \{r_{j+1}\} \rightarrow \{r_{j+1}\} - \theta_k\{r_k\} \tag{11.76a}$$

$$\{w_{j+1}\} \rightarrow \{w_{j+1}\} - \phi_k\{w_k\}, \quad \{s_{j+1}\} \rightarrow \{s_{j+1}\} - \phi_k\{s_k\} \tag{11.76b}$$

The value of ε_0 is dependent on the machine precision of the computer employed. Usually, a value of 10^{-8} has been found to work well. A much smaller value of ε_0 will yield Lanczos vectors with a higher level of biorthogonality. However, the algorithm also depends on the finite precision of the machine to capture repeated eigenvalues that may be present in a problem, as described in reference [195]. Therefore, the value of ε_0 should be chosen at least slightly higher than the machine precision in order for the repeated eigenvalues to converge. The value of $\varepsilon_0 = 10^{-8}$ is based on the numerical experimentation such that a repeated eigenvalue converged within five to ten recursion steps after the first has converged. When there are complex eigenvalues in a problem, they converged in conjugate pairs.

Eigenvalue search strategy

The eigenvalues in a specified magnitude range are searched by introducing a shift σ equal to the lower limit of the eigenvalue magnitudes sought. The search procedure employed for the quadratic eigenproblem here is exactly same as that used for the non-symmetric generalized eigenproblem described in Section 11.3.3. In the quadratic eigenproblem, however occasional break down of the recursion may occur due to the possibility of the sum of the scalar products resulting in a null value for the normalization coefficient Δ_{j+1} computed in equation (11.63). As can be seen in equations (11.64) through (11.66), this will result in a non-definable $(j+1)$st Lanczos vectors. In this event the remedy is to restart the recursion, after saving the converged eigenvalues, with a new set of starting vectors. This occurrence is, however, rare. If $|\Delta_{j+1}|$ drops close to zero, there are two possibilities: (a) the aforementioned breakdown has occurred, and (b) the eigenvalues have all converged in the range of $|\sigma| < |\lambda_i| < |\lambda_k|$, for $k < j$, and the magnitude of the next eigenvalue, λ_{k+1}, is much higher than that of λ_k, often by more than two orders of magnitude. Under such conditions, the best approach is to restart the recursion employing a new shift of magnitude slightly lower than that of the largest converged eigenvalue in the terminated recursion.

11.7.4. Quadratic eigenvalue problem examples

Two examples are presented here to demonstrate the application of the Lanczos recursion for quadratic eigenvalue problems. First one is an input matrix problem where three randomly generated matrices from known eigenvalues form the $[A]$, $[B]$ and $[C]$ matrices of the quadratic eigenvalue problem. The second one is a simply supported beam modeled using the FEM with damping specified to absorb energy. For additional examples showing the application of the method to Boundary Element eigenvalue problem the reader is directed to Chapter 10, examples 4, 5, and 6 where acoustic cavity eigenvalue problems have been considered.

Example 11.3

This is a small size (6×6) input matrix problem showing the validation of the Lanczos recursion scheme for the quadratic eigenvalue problem. The matrices $[A]$, $[C]$, and $[B]$ are generated from a set of known complex eigenvalues $\lambda_i = \sigma_i - \mathrm{j}\omega_i$, $\lambda_{i+1} = \sigma_i + \mathrm{j}\omega_i$ where $\mathrm{j} = \sqrt{-1}$. First, three diagonal matrices are formed as follows: $[A_\mathrm{d}] = [\sigma_i^2 + \omega_i^2]$, $[C_\mathrm{d}] = [-2\sigma_i]$ and $[B_\mathrm{d}] = [I]$, $i = 1, 2, \ldots, n$ $(n = 6)$. It can be seen that the eigenvalues of the diagonal quadratic eigenvalue problem $[A_\mathrm{d}]\{x_i\} + \lambda_i[C_\mathrm{d}]\{x_i\} + \lambda_i^2[B_\mathrm{d}]\{x_i\} = 0$ are given by $\lambda_i = \sigma_i \pm \mathrm{j}\omega_i$. Then the diagonal matrices are pre-multiplied by an arbitrary

Table 11.4. Input matrices of Example 11.3.

[A] Matrix					
324	−2190	23160	10864415	−4182340	109027740
72	−2482	−1930	4205580	13383488	−7268516
1152	657	13510	−8411160	14219956	32708322
612	1314	−13510	−2803720	12547020	85405063
900	−2409	−52110	18224180	−11710552	61782386
1728	−2263	88780	−2102790	3764106	−47245354
[C] Matrix					
−108	180	−216	−62	−940	161760
24	204	18	−24	3008	−10784
−384	−54	−126	48	3196	48528
−204	−108	126	16	2820	126712
−300	198	486	−104	−2632	91664
−576	186	−828	12	846	−70096
[B] Matrix					
9	−30	12	31	−10	60
2	−34	−1	12	32	−4
32	9	7	−24	34	18
17	18	−7	−8	30	47
25	−33	−27	52	−28	34
48	−31	46	−6	9	−26

Table 11.5. Eigenvalues of Example 11.3.

	Eigenvalues		
No.	Known	Computed	
1	6 + 0J	6.0004588	0.99258436E−06J
2	6 − 0J	5.9995408	−0.98979812E−06J
3	3 + 8J	2.9999997	8.0000007J
4	3 − 8J	2.9999997	−8.0000007J
5	9 + 43J	9.0000022	42.999999J
6	9 − 43J	9.0000022	−42.999998J
7	1 + 592J	1.0000667	591.99991J
8	1 − 592J	1.0000655	−591.99991J
9	−47 + 645J	−46.999961	644.99963J
10	−47 − 645J	−46.999959	−644.99963J
11	−1348 + 5J	−1348.0007	5.0045941J
12	−1348 − 5J	−1348.0007	−5.0045945J

matrix $[E]$ of size 6×6 whose elements are randomly filled. The quadratic eigenproblem formed by the product matrices $[A] = [E][A_d]$, $[C] = [E][C_d]$, $[B] = [E][B_d]$ have the same eigenpairs as that of the diagonal quadratic eigenvalue problem. The input matrices generated by this process are given in Table 11.4.

The computed eigenvalues of this quadratic eigenvalue problem are listed in Table 11.5 along with the known eigenvalues of problem. As we know, the eigenvalues of the quadratic eigenproblem appear in complex conjugate pairs. For the 6×6

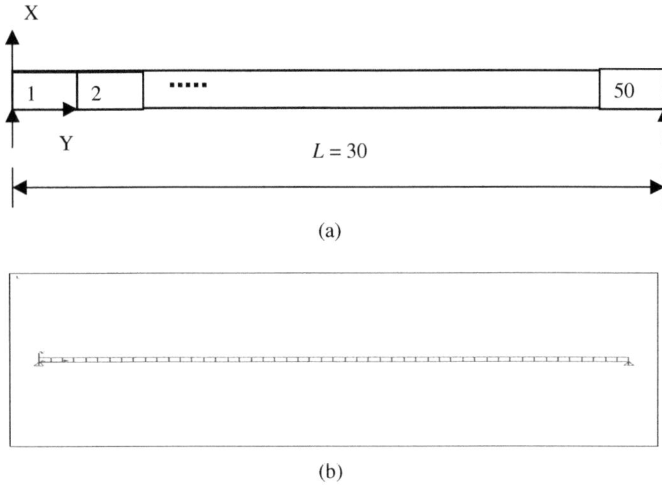

Figure 11.2. Simply supported beam of square cross section (Young's modulus $E = 12E10$, density $\rho = 1E4$, area $A = 1$, moment of inertia $I = 1/12$, proportional damping coefficients: $\alpha = 0.12$, $\beta = 0.003$). (a) Simply supported beam; (b) Finite element discretization.

system all of the 6 conjugate pairs of eigenvalues were computed by taking $m = 12$ Lanczos steps. In spite of the wide spread of the eigenvalue magnitude spectrum, the eigenvalues converged and are in good agreement with the actual eigenvalues of the problem.

Example 11.4

A simply supported beam of square cross section, shown in Figure 11.2, is discretized by finite elements to compute its damped system resonant frequencies. The damping matrix that accounts for the dissipation of energy is formed by considering proportional damping expression given by $[C] = \alpha[M] + \beta[K]$. Here, α and β are the scalar coefficients that bring a fraction of the mass and stiffness matrices, $[K]$ and $[M]$, respectively, of the beam into $[C]$. The resulting quadratic eigenvalue problem of the beam is given by

$$[K]\{x_i\} + \lambda_i[C]\{x_i\} + \lambda_i^2[M]\{x_i\} = 0 \tag{11.77}$$

The beam was divided into 50 finite elements along its length employing the ANSYS® program and the damped system frequencies computed using the Lanczos recursion scheme. The computed eigenvalues of the damped system eigenproblem are presented in Table 11.6. The eigenfrequencies of the beam are given by the following analytical expression [197].

$$\frac{\lambda_n^2 + \alpha\lambda_n}{1 + \beta\lambda_n} = -\left(\frac{n\pi}{L}\right)^4 \frac{EI}{\rho A}, \quad n = 1, 2, \ldots, \infty \tag{11.78}$$

where λ_n are the circular eigenfrequencies of the beam. The material properties of the beam, Young's modulus and the density, are shown in Figure 11.2. The cross section area and moment of inertia are given by A and I, respectively. In Table 11.6 the first ten frequencies computed are listed along with the analytical frequencies. The agreement of the computed values with the analytical frequencies is quite good. The

Table 11.6. Eigenfrequencies of the beam with proportional damping.

Mode	Frequency (Hz)			
	Analytical		Computed	
1	−0.38258801E−01	+1.7449094 J	−0.38231453E−01	+1.7440785 J
2	−0.38258801E−01	−1.7449094 J	−0.38231453E−01	−1.7440785 J
3	−0.46890141	+6.9655503 J	−0.46721040	+6.9527743 J
4	−0.46890141	−6.9655503 J	−0.46721040	−6.9527743 J
5	−2.3350194	+15.533437 J	−2.3159626	+15.471081 J
6	−2.3350194	−15.533437 J	−2.3159626	−15.471081 J
7	−7.3591831	+26.938125 J	−7.2530238	+26.757246 J
8	−7.3591831	−26.938125 J	−7.2530238	−26.757246 J
9	−17.952991	+39.768680 J	−17.551764	+39.410940 J
10	−17.952991	−39.768680 J	−17.551764	−39.410940 J

slight differences are partly attributable to the discretization error inherent in the finite element discretization of the continuum.

11.8. Summary statements on eigenvalue computation algorithms

In this chapter we presented the tools needed to efficiently solve the boundary element eigenvalue problems. It turns out that the algebraic eigenvalue problems developed in Chapters 7 through 10 can all be handled using one of the Lanczos computational schemes presented in this chapter. In the literature, in addition to the Lanczos two-sided recursion schemes, the Arnoldi's method is found to be used for non-symmetric eigenvalue problems. While the two methods share the common Krylov sequence of vectors, they differ in the resulting subspace matrix. The Lanczos two-sided recursion is attractive due to the fact that the resulting subspace matrix is tridiagonal whose eigenvalues can be extracted quite efficiently. Furthermore, the Lanczos algorithm has been quite widely employed in commercial finite element software for the symmetric problem, and thus the incorporation of the non-symmetric schemes would involve extension of the implementations to the non-symmetric problem.

In addition, symmetric boundary element formulations have been actively pursued to make the method more attractive for large practical problems. The Lanczos schemes developed in this chapter can be quite easily specialized for symmetric eigenproblems.

Chapter 12

Discussion and Future Research

12.1. Discussion on boundary element eigenvalue methodologies

The evolutionary history of the development of eigenvalue formulations, based on the boundary element method (BEM), is presented in this book. The material is presented in a chronologically organized manner, starting from the early Determinant Search Method (DSM) to the most recently proposed techniques such as the Multiple Reciprocity Method (MRM), Series Expansion Method (SEM), etc. An alternative approach would have been to organize the materials based upon the fields of application such as two-dimensional (2-D) elasticity, three-dimensional (3-D) elasticity, plates and shells, 2-D acoustics and so on. This would, however, have turned us away from our stated goal to focus our attention on the development of eigenvalue formulation itself. We chose the chronological approach over the application-driven approach because the BEM based eigenvalue formulations are still in an evolving stage and more research will be required in a number of related areas to improve the degree of maturity of the technique. The book will hopefully encourage researchers to continue work on the subject.

We have reviewed and provided detailed eigenvalue formulations for acoustics, elasticity and plate vibrations. All the formulations, proposed so far in the literature to solve BE eigenvalue problems such as Determinant Search Method, enhanced Determinant Search Method, Internal Cell Method (ICM), Dual Reciprocity Method (DRM), Particular Integral Method (PIM) and its variation, Multiple Reciprocity Method (MRM), Series Expansion Method (SEM) with matrix augmentation and so on, are discussed in the book. Wherever appropriate, the advantages and disadvantages of each method are also recorded in the relevant chapters. In the next section we will briefly compare BE eigenvalue formulations with finite element (FE) eigenvalue formulation. A few topics related to BE eigenvalue formulations are not covered in the book. These topics are enumerated in Section 12.3. The final section is dedicated to the discussion of possible future research that can be undertaken in order to improve the performance of BE eigensolution methodology and to extend its capability.

The book is written such that a beginner on the subject of BEM can learn the basics of the method and then gradually get into the discussion of eigenvalue formulations. The basics of BEM are developed using simple potential problems to make it easier for the reader to enter into the BEM subject matter without difficulty. All the important details of basic BE formulation are worked out in Chapters 2 through 4. A sufficient number of example problems are also presented in these chapters in order to demonstrate the application of the BE formulations developed there.

On the surface the BE-based eigenanalysis will appear to be a very cumbersome and awkward proposition. The free-field fundamental solutions used in the BE formulations are transcendental in form, involving trigonometric, logarithmic and exponential functions. These functions contain the frequency parameters implicitly in them. Thus, in a direct application of the boundary integral principle, the frequency parameters cannot be factored out of the integrals. This forces one to re-compute the BE matrices at every iteration and to use DSM, which is awkward and inaccurate. The use of the fundamental solution for the complete governing differential equation is motivated by the classical boundary integral technique used till middle nineteen seventies. The usage was mostly confined to indirect boundary integral methods with uniform distribution of source densities over each discretized boundary segment. The Green's integral transformation formulas were mostly used to form the boundary integrals, leading to compact, global and rather inflexible formulations. There was no concept of separate mass-like or stiffness-like matrices in these formulations. All were lumped together in a mixed formulation. The method was perceived to be disconnected from the domain type numerical methodologies such as the finite difference or finite element techniques.

Starting from mid-1970s, the researchers discovered the similarities between boundary integral technique and the domain methods. This led to the popularity of the "direct" BEM, and the use of *isoparametric shape functions* in the description of the physical variables of the problem, just like the finite element methods (FEMs). The researchers, trained mainly in the domain methods, started using the weighted residual and energy principles, as they used in the domain methods, to formulate discretized boundary element equations.

In an attempt to form a separate mass-like matrix, as in the FEM, the inertia term of the governing differential equation was treated separately. The remaining terms of the differential equations allowed the use of time-independent "static fundamental solution," which led to the formation of stiffness-like co-efficient matrix. The process lent itself to elegant algebraic eigenvalue formulation. Initially the domain was divided into cells (ICM) to integrate the mass matrix term. Later the DRMs and PIMs subjected the inertia domain term to further integral transformation and thus eliminated the need to use domain cells to form mass matrix, leading to boundary-only BE algebraic eigenvalue formulations.

DRM and PIM or its variants such as the Fictitious Function Method (FFM) or the Polynomial-based PIM are powerful and elegant BE algebraic eigenvalue formulations and can be implemented in commercial codes for routine free vibration analysis. The mechanism of inserting internal points or zoning may be automated in the code as an integral part of the eigenvalue solution procedure.

The MRM combined with an appropriate eigensolver or MRM with matrix augmentation combined with an appropriate eigensolver can also be implemented in the commercial codes. However, before such implementation takes place, the process, especially the eigensolver, must be fine-tuned and automated to exploit the characteristics of the coefficient and mass-like matrices, generated using these formulations.

12.2. Comparison of eigenanalysis using BEM and FEM

In BEM, the discretization is confined to the boundary alone, which results in significant reductions in the preprocessing efforts and leads to smaller problem sizes compared to those in the FE formulation. The generation of adequate FE meshes for complex 3-D problems is still not fully automated, except to fill the volumes with large number of tetrahedron elements. Even if FE mesh generation is fully automated, an enormous amount of data needs to be handled for pre- and post-processing functions for complex 3-D problems. On the contrary, BEA will stop at the discretization of the surface of the 3-D domain, thereby reducing mesh generation efforts and data creation.

The BE [A] and [B] matrices are fully populated and unsymmetric, which forces one to find special eigensolvers suitable for unsymmetric matrices. Furthermore, BE eigenvalue problems (DRM and PIM-type formulations), especially in acoustic resonant frequency computations, will require additional field collocation points for accurate answers, thereby increasing the size of the matrices.

Summing up, the overall matrix sizes in the BE eigenvalue formulations are smaller than those in the FE formulation. For a coupled fluid–structure interaction eigenvalue problem, where the fluid is contained in a structure, the FE formulation also produces unsymmetric mass and stiffness matrices. In this case, BE cigcnvaluc formulation will obviously be more efficient than FE formulation.

12.3. Topics not covered in the book

It was mentioned in Section 12.1 that we covered all the BE eigenvalue formulations appearing in the literature and we discussed all relevant issues relating to BE eigenvalue analysis. However, there are a few aspects, related to BE eigenvalue analysis, that were not considered. The important ones are: axisymmetric BE eigenvalue problem in elasticity, the so-called dual MRM and the use of residual method or singular value decomposition technique to filter out spurious frequencies, symmetric BE eigenvalue formulation, adaptive BE eigenvalue analysis [147] and detailed development of BE eigenvalue formulation for bi-harmonic problem [146]. A couple of these rather important topics are briefly touched upon below.

Axisymmetric BE eigenvalue problem in elasticity

Axisymmetric solids can be analyzed as full-blown 3-D bodies. By taking the axisymmetry into consideration one can solve the problem essentially in 2-D. It means that axisymmetric bodies can be discretized using one-dimensional (1-D) BE line segments [125, 126, 129]. So far as the BE eigenvalue formulation is concerned, we can utilize DRM or PIM, developed in Chapter 8. In this case, the static fundamental solution for axisymmetric bodies will have to be used. This will require evaluation of elliptic integrals. See Section 4.3 for details on BE axisymmetric formulations.

Symmetric BE formulation

In earlier boundary integral applications, the integral transformation methods such as the Gauss's divergence theorem were used to formulate the boundary element equations. After discretization, this approach naturally led to unsymmetric system matrices. Later employment of weighted residual techniques also produced unsymmetric matrices. Starting in the mid-1970s some researchers utilized variational and energy principles, especially in the context of coupling domain methods with boundary elements [e.g., 65], and derived symmetric boundary element matrices.

Loosely speaking, two different boundary element approaches emerged. One group pioneered the development of modern BEM, popularized the direct BEM, opened up the possibility of the use of higher order isoparametric shape functions and wrote boundary element codes from scratch and not as an appendage to existing domain method codes. The members of this research group accepted the non-symmetry of the boundary element system matrices naturally as a given fact and implemented solvers in their computer programs suited to deal with unsymmetric and fully populated matrices. Note that because the co-efficient and mass-like matrices are unsymmetric, the Maxwell–Betti reciprocal theorem is violated, even though the mathematical problem is self-adjoint.

The members of the second group of researchers were already well-established in domain methods such as the FEM. They wanted to investigate boundary element as a tool, which could provide accurate boundary conditions in the solution of far-field problems by their domain method codes. Here the preservation of symmetry and bandedness of the system matrices produced by the domain methods was important. Encouraged by the outcome of the early research work along this line, boundary element researchers subsequently employed variational and energy principles to formulate time-dependent symmetric boundary element equations including eigenformulation equations.

In order to formulate algebraic symmetric boundary element elasticity eigenvalue problem, we can, for example, write a functional in terms of boundary displacement, boundary traction and domain displacement [148]. Assume that these variables are independent of one another. We can then:

(i) Discretize the boundary of the domain;
(ii) Express the boundary variables by their nodal values;
(iii) Approximate the domain displacements with global shape functions (GSFs) or with a linear combination of static fundamental solutions; and
(iv) Compute variations of the functional with respect to the three independent variables.

The domain integral representing the inertia term is transformed into boundary integral using DRM. The stiffness and mass matrices produced in this fashion are symmetric and positive definite.

Let us note here that the symmetric boundary element formulations arising out of variational or energy principles are said to be wrong by some researchers [e.g., 71]. Suppose $[K]$ is an unsymmetric global boundary element system matrix and suppose it is symmetrized as $[K'] = \frac{1}{2}([K] + [K]^{\mathrm{T}})$. It is shown in reference 71 (see Chapter 13) that the energy approach produces a system matrix which is equivalent to $[K']$. It has been shown elsewhere [198] that when boundary element is coupled with domain method such as FEM, the boundary element coefficient matrix can be symmetrized without significant loss of accuracy in solutions.

12.4. Future research on BEM eigenanalysis

The subject of boundary element eigenvalue formulation is still evolving. The efforts of successful commercialization of BE eigenanalysis for routine use by common engineers are ongoing. Researchers are continuing investigations to answer questions on various issues relating to BE eigenanalysis. Below we list a few of these issues that will certainly require further research.

MRM formulation

The MRM eigenvalue formulation or equivalently the formulation that uses Helmholtz fundamental solution with expansion of matrices in terms of the wave numbers eliminates the need for internal collocation points. However, in this case we need to take a step backward and use either a DSM or a Newton–Raphson iteration procedure, both of which are difficult to use and inefficient in a general purpose environment. Although the matrix augmentation technique allows one to use generalized eigensolvers readily available in standard eigensolver packages, the matrix size becomes significantly large. Consequently, except for the differences in meshing efforts, this procedure defeats the purpose of using boundary-only discretization.

Additionally, the matrix augmentation procedure is reported to introduce spurious eigenmodes into the solution, which should be filtered out using residual method or singular value decomposition technique [155] so that the extracted eigenvalues are reliable. Furthermore, the generalized eigensolvers available in standard packages do not currently exploit the fact that the augmented matrices are very sparse and contain numerous null and identity sub-matrices.

Zoned BEM/internal collocation points

It was discussed in Chapter 9 that we need to break up the domain into zones or put additional field collocation points to improve solution accuracy in DRM and PIM formulations. This is especially true for chunky-shaped acoustic cavities. Original proponents of DRM such as Nardini and Brebbia realized, as early as 1982 [116], that internal collocation points will be required in order to improve the accuracy of the eigensolution, especially higher modes of free vibrations. Kanarachos and Provatidis [122] later noted that the refinement of boundary discretization after certain point does not improve the quality of eigensolution. They concluded: the set of boundary functions used to approximate the inertia term is not complete and consequently the corresponding BE solution will not converge. In order to complete the set, field sources will have to be introduced. They designated the complete set as the field or Poisson-adjusted functional set. The issue involving the selection of appropriate shape functions for the approximation of the inertia term is still being debated [199–201].

Regarding zoning and internal collocation points, no clear guideline exists in the current literature on how to break up the domain into different zones or where and how many internal collocation points should be placed. Such a guideline, based on extensive parametric studies, ought to be established before BE eigenvalue analysis codes using DRM/PIM can be developed for general-purpose use. The best scenario would be to establish a procedure, as part of generating boundary mesh, to automatically create an optimal network of zones or put an optimal number of strategically distributed internal collocation points. If this is achieved, BE eigenvalue analysis using DRM/PIM type formulation would have a better chance to compete against the FE eigenformulation as an effective numerical tool.

Submerged structures

When a structure is submerged in a fluid of infinite extent, e.g., a submarine submerged in seawater, the vibration characteristics of the structure change dramatically.

Although the problem can be addressed by FEM, it is far from being efficient, as a large fluid domain needs to be included in the FE model in order to solve the problem with a reasonable accuracy. To our knowledge, this problem has not yet been addressed in the context of BEM. Especially, how to compute an accurate fluid mass matrix in this situation is still an open question that needs to be answered.

Symmetric BE eigenvalue formulation

In Section 12.3 above, we briefly described the ongoing debate on this topic. The issue has not yet been resolved. There is scope to conduct further investigation into both arenas: (a) direct development of symmetric matrices for BE algebraic eigenvalue problem and (b) possible standardization of methods of symmetrizing unsymmetric BE matrices.

References

1. I. Fredholm. Sur une classe d'equations fonctionelles. *Acta Mathematica, Sweden* **27**, 365–390 (1903).
2. E. Trefftz. Über die Kontraktion Kreisförmiger Flüssigkeitsstrahlen. *Zeitschrift für Mathematik und Physik* **64**, 34–61 (1917).
3. W. Prager. Die Druckverteilung an Körpern in Ebener Potentialströmung. *Physikalische Zeitschrift* **29**, 865–869 (1928).
4. O.D. Kellog. *Foundations of Potential Theory*. Dover, New York (1953). Originally published by J. Springer, Berlin (1929).
5. C. Somigliana. Sopra l'equilibrio di un corpo elastico isotropo. *Il Nuovo* 17–19 (1886).
6. N.I. Muskhelishvili. *Some Basic Problems of the Mathematical Theory of Elasticity*. Noordhoff, Groningen (1953).
7. N.I. Muskhelishvili. *Singular Integral Equations: Boundary Problems of Function Theory and Their Applications to Mathematical Physics*, 2nd edition. Dover Publications, New York (July 1992).
8. S.G. Mikhlin. *Integral Equations*. Pergamon Press, Oxford (1957).
9. A.M.O. Smith and J. Pierce. Exact solution of the Neumann problem, calculation of non-circulatory plane and axially symmetric flows about or within arbitrary boundaries. In: R.M. Haythornthwaite (Ed.), *Proceedings of the 3rd US National Congress of Applied Mechanics*. American Society of Mechanical Engineers, New York, pp. 807–815 (1958).
10. M.B. Friedman and R.P. Shaw. Diffraction of pulses by cylindrical obstacles of arbitrary cross section. *Journal of Applied Mechanics* **29**, 40–46 (1962).
11. M.B. Friedman and R.P. Shaw. Diffraction of pulses by deformable cylindrical obstacles of arbitrary cross section. In: R.M. Rosenberg (Ed.), *Proceedings of the 4th US National Congress of Applied Mechanics*. American Society of Mechanical Engineers, New York, pp. 371–379 (1962).
12. R.P. Shaw. Diffraction of acoustic pulses by obstacles of arbitrary shape with a Robin boundary condition—Part A, *Journal of the Acoustical Society of America* **41**, 855–859 (1962).
13. R.P. Banaugh and Goldsmith. Diffraction of steady acoustic waves by surfaces of arbitrary shape. *Journal of the Acoustical Society of America* **35**, 1590–1601 (1963).
14. J.L. Hess. Calculation of potential flow about bodies of revolution having axes perpendicular to the free stream direction. *Journal of the Aerospace Sciences* **29**, 726–742 (1962).
15. J.L. Hess and A.M.O. Smith. Calculation of nonlifting potential flow about arbitrary three-dimension bodies. *Journal of Ship Research* **8**, 22–44 (1964).
16. M.A. Jawson. Integral equation methods in potential theory: I. *Proceedings of the Royal Society of London* (A) **275**, 23–32 (1963).
17. G.T. Symm. Integral equation methods in potential theory: II. *Proceedings of the Royal Society of London* (A) **275**, 33–46 (1963).
18. M.A. Jawson and A.R.S. Ponter. An integral equation solution of the torsion problem. *Proceedings of the Royal Society of London* (A) **273**, 237–246 (1963).

19. C.E. Massonnet. Numerical use of integral procedures, In: O.C. Zienkiewicz and G.S. Holister (Eds.), *Stress Analysis*. John Wiley and Sons, London, pp. 198–235 (1965).
20. V.D. Kupradze. *Potential Methods in the Theory of Elasticity*. Israel Program for Scientific Translations, Jerusalem (1965).
21. S.G. Mikhlin. *Approximate Solutions of Differential and Integral Equations*. Pergamon Press, Oxford (1965).
22. S.G. Mikhlin. *Multidimensional Singular Integrals and Integral Equations*. Pergamon Press, Oxford (1965).
23. M.A. Jawson, M. Maiti and G.T. Symm. Numerical biharmonic analysis and some applications. *International Journal of Solids and Structures* **3**, 309–332 (1967).
24. K. Rim and A.S. Henry. *An integral equation method in plane elasticity*. NASA Contractor Report CR-779 (1967).
25. F.J. Rizzo. An integral equation approach to boundary value problems of classical elastostatics. *Quarterly of Applied Mathematics* **25**, 83–95 (1967).
26. E.R.A. Oliveira. Plane stress analysis by a general integral method. *Journal of the Engineering Mechanics Division, Proceedings of the ASCE* **94**, (EM 1), 79–101 (1968).
27. T.A. Cruise and F.J. Rizzo. A direct formulation and numerical solution of the general transient elastodynamic problem I. *Journal of Mathematical Analysis and Applications* **22**, 244–259 (1968).
28. T.A. Cruise. A direct formulation and numerical solution of the general transient elastodynamic problem II. *Journal of Mathematical Analysis and Applications* **22**, 341–355 (1968).
29. T.A. Cruise. Numerical solution in three-dimensional elastostatics. *International Journal of Solids and Structures* **5**, 1259–1274 (1969).
30. M.A. Jawson and M. Maiti. An integral equation formulation of plate bending problems. *Journal of Engineering Mathematics* **2**, 83–93 (1968).
31. D.A. Newton and H. Tottenham. Boundary value problems in thin shallow shells of arbitrary plan form. *Journal of Engineering Mathematics* **2**, 211–224 (1968).
32. D.J. Forbes and A.R. Robinson. Numerical analysis of elastic plates and shallow shells by an integral equation method. *Structural Research Series. Report No. 345*. University of Illinois, Urbana, Illinois (1969).
33. R.F. Harrington, K. Kontoppidan, P. Abrahamsen and N.C. Albertsen. Computation of Laplacian potentials by an equivalent-source method, *Proceedings of the Institution of Electrical Engineers* **116**, 1715–1720 (1969).
34. R. Butterfield and P.K. Banerjee. The problem of pile reinforced half space. *Geotechnique* **20**, 100–103 (1970).
35. R. Butterfield and P.K. Banerjee. The elastic analysis of compressible piles and pile groups. *Geotechnique* **21**, 43–60 (1971).
36. F.J. Rizzo and D.J. Shippy. A method of solution for certain problems of transient heat conduction. *American Institute of Aeronautics and Astronautics Journal* **8**, 2004–2009 (1970).
37. F.J. Rizzo and D.J. Shippy. An application of the correspondence principle of linear viscoelasticity. *SIAM Journal of Applied Mathematics* **21**, 321–330 (1971).
38. T.A. Cruise and W. Van Buren. Three-dimensional elastic stress analysis of a fractured specimen with an edge crack, *International Journal of Fracture Mechanics* **7**, 1–15 (1971).
39. J.L. Swedlow and T.A. Cruise. Formulation of boundary integral equations for three-dimensional elastoplastic flow. *International Journal of Solids and Structures* **7**, 1673–1681 (1971).
40. R.P. Shaw. Methods of solution for water wave scattering problems. In: C. Bretschneider (Ed.), *Topics in Ocean Engineering* II. Gulf Publishing Company, Houston, pp. 180–198 (1970).
41. J.J. Lee. Wave induced oscillation in harbors of arbitrary geometry. *Journal of Fluid Mechanics* **45**, 375–394 (1971).
42. P. Silvester and M.S. Hsieh. Finite element solution of 2-dimensional exterior field problems. *Proceedings of the Institution Electrical Engineers* **118**, 1743–1747 (1971).
43. B.H. McDonald and A. Wexler. Finite-element solution of unbounded field problems. *IEEE Transactions on Microwave Theory and Techniques* **MTT-20**, 841–847 (1972).

44. R. Benjumea and D.L. Sikarskie. On the solutions of plane, orthotopic elasticity problems by an integral equation method. *Journal of Applied Mechanics* **39**, 801–808 (1972).
45. T.A. Cruise. Application of the boundary integral equation method to three-dimensional stress analysis. *Computers and Structures* **3**, 509–527 (1973).
46. S.L. Crouch. Two-dimensional analysis of near-surface single seam extraction. *International Journal of Rock Mechanics and Mining Sciences & Geomechanics Abstracts* **10**, 85–96 (1973).
47. P.K. Banerjee. Integral equation methods for analysis of piecewise nonhomogeneous three-dimensional elastic solids of arbitrary shape. *International Journal of Mechanical Sciences* **18**, 293–303 (1976).
48. S.L. Crouch. Solution of plane elasticity problems by the displacement discontinuity method, I and II. *International Journal for Numerical Methods in Engineering* **10**, 301–343 (1976).
49. A. Mendelson. Boundary-integral methods in elasticity and plasticity. *NASA Technical Note* TN D-7418, Washington DC (1973).
50. F. Erdogan, G.D. Gupta and T.S. Cook. Numerical solution of integral equations. In: G.C. Sih (Ed.) *Mechanics of Fracture, Vol. 1, Methods of Analysis and Solutions of Crack Problems*. Noordhoff Publishing Company, Leiden, pp. 368–425 (1973).
51. P.S. Theocaris and N.I. Ioakimidis. Numerical integration methods for the solution of singular integral equations. *Quarterly of Applied Mathematics* **35**, 173–183 (1977).
52. P. Riccardella. *An Implementation of the Boundary Integral Technique for Plane Problems of Elasticity and Elastoplasticity*. Ph.D. Thesis, Carnegie Mellon University, Pittsburgh, Pennsylvania (1973).
53. V. Kumar and S. Mukherjee. A boundary integral equation formulation for time-dependent inelastic deformation in metals. *International Journal of Mechanical Sciences* **19**, 713–724 (1977).
54. J.C. Wu and J.F. Thompson. Numerical solution of time-dependent incompressible Navier–Stokes equations using an integrodifferential formulation. *Computers and Fluids* **1**, 197–215 (1973).
55. J.C. Wu, A.H. Spring and N.L. Sankar. A flow field segmentation method for numerical solution of viscous flow problems. *Lecture Notes in Physics*. Vol. 35. Springer-Verlag, Berlin, pp. 452–457 (1974).
56. Y. Niwa, S. Kobayashi and T. Fukui. An application of the integral equation method to seepage problems. *Theoretical and Applied Mechanics* **24**, 479–486 (1976).
57. J.A. Liggett. Locations of free surface in porous media. *Journal of the Hydraulics Division, Proceedings of the ASCE* **103**, HY 4, 353–365 (1977).
58. F.J. Rizzo and D.J. Shippy. An advanced boundary integral equation method for three-dimensional thermoelasticity. *International Journal for Numerical Methods in Engineering* **11**, 1753–1768 (1977).
59. T.A. Cruise and F.J. Rizzo (Eds.). *Boundary Integral Equation Method: Computational Applications in Applied Mechanics*. AMD-11, American Society of Mechanical Engineers, New York (1975).
60. P.K. Banerjee and Butterfield. Boundary element methods in geomechanics. In: G. Gudehus (Ed.), *Finite Elements in Geomechanics*. John Wiley and Sons, London, pp. 529–570 (1977).
61. C.A. Brebbia and J. Dominguez. Boundary element methods for potential problems. *Applied Mathematical Modelling* **1**, 372–378 (1977).
62. C.A. Brebbia. Approximate methods in mathematical modelling. In: X.J.R. Avula (Ed.), *Proceedings of the 1st International Conference on Mathematical Modelling*. Department of Engineering Mechanics, University of Missouri-Rolla, pp. 43–57 (1977).
63. C.A. Brebbia. *The Boundary Element Method for Engineers*. Pentech Press, London (1978).
64. M.A. Jawson and G.T. Symm. *Integral Equation Methods in Potential Theory and Elastostatics*. Academic Press, London (1977).
65. O.C. Zienkiewicz, D.W. Kelly and P. Bettess. The coupling of the finite element method and boundary solution procedures. *International Journal for Numerical Methods in Engineering* **11**, 355–375 (1977).
66. O.C. Zienkiewicz. *The Finite Element Method*, 3rd ed. McGraw-Hill, London (1977).

67. S.N. Atluri and J.J. Grannell. Boundary element methods (BEM) and combination of BEM-FEM. *Report No. GIT-ESM-SA-78-16*. Center for the Advancement of Computational Mechanics, Georgia Institute of Technology, Atlanta, Georgia (1978).
68. C.A. Brebbia and R. Butterfield. Formal equivalence of direct and indirect boundary element methods. *Applied Mathematical Modelling* **2**, 132–134 (1978).
69. C.A. Brebbia and Walker. *Boundary Element Techniques in Engineering*. Newnes-Butterworths, London (1980).
70. P.K. Banerjee and R. Butterfield. *Boundary Element Methods in Engineering Science*. McGraw-Hill, London (1981).
71. C.A. Brebbia, J.C.F. Telles and L.C. Wrobel. *Boundary Element Techniques*. Springer-Verlag, Berlin (1984).
72. S. Mukherjee. *Boundary Element Methods in Creep and Fracture*. Applied Science, London (1982).
73. V.Z. Parton and P.I. Perlin. *Integral Equations in Elasticity*. Mir Publishers, Moscow (1982).
74. S.L. Crouch and A.M. Starfield. *Boundary Element Methods in Solid Mechanics*. George Allen and Unwin, London (1983).
75. J.A. Liggett and P.L.-F. Liu. *The Boundary Integral Equation Method for Porous Media Flow*. George Allen and Unwin, London (1983).
76. J.C.F. Telles. *The Boundary Element Method Applied to Inelastic Problems*. Springer-Verlag, Berlin (1983).
77. W.S. Venturini. *Boundary Element Method in Geomechanics*. Springer-Verlag, Berlin (1983).
78. T.V. Hromadka II. *The Complex Variable Boundary Element Method*. Springer-Verlag, Berlin (1984).
79. D.B. Ingham and M.A. Kelmanson. *Boundary Integral Equation Analysis of Singular Potential and Biharmonic Problems*. Springer-Verlag, Berlin (1984).
80. P.K. Banerjee and R. Butterfield (Eds.). *Developments in Boundary Element Methods-1*. Applied Science, London (1979).
81. C.A. Brebbia (Ed.). *Progress in Boundary Element Methods*. Vol. 1. Pentech Press, London (1981).
82. P.K. Banerjee and R.P. Shaw (Eds.). *Developments in Boundary Element Methods-2*. Applied Science, London (1982).
83. C.A. Brebbia (Ed.). *Progress in Boundary Element Methods*. Vol. 2. Pentech Press, London (1983).
84. P.K. Banerjee and S. Mukherjee (Eds.). *Developments in Boundary Elements-3*. Elsevier Applied Science, London (1984).
85. C.A. Brebbia (Ed.). *Topics in Boundary Element Research*. Vol. 1. Springer-Verlag, Berlin (1984).
86. C.A. Brebbia (Ed.). *Topics in Boundary Element Research*. Vol. 2. Springer-Verlag, Berlin (1985).
87. P.K. Banerjee and J.O. Watson (Eds.). *Developments in Boundary Elements-4*. Elsevier Applied Science, London (1986).
88. D.E. Beskos (Ed.). *Boundary Element Methods in Mechanics*. Elsevier Science Publishers, Amsterdam (1987).
89. C.A. Brebbia (Ed.). *New Developments in Boundary Element Methods*. CML Publication, Southampton (1980).
90. C.A. Brebbia (Ed.). *Boundary Element Methods*. Springer-Verlag, Berlin (1981).
91. C.A. Brebbia (Ed.). *Boundary Element Methods in Engineering*. Springer-Verlag, Berlin (1982).
92. C.A. Brebbia, T. Futagami and M. Tanaka (Eds.). *Boundary Elements*. Springer-Verlag, Berlin (1983).
93. C.A. Brebbia (Ed.). *Boundary Elements* IV. Springer-Verlag, Berlin (1984).
94. C.A. Brebbia, R.P. Shaw, M. Tanaka and M.H. Aliabadi (Eds.). *Engineering Analysis with Boundary Elements*. Elsevier Applied Science, Oxford, U.K.
95. M.H. Aliabadi, C.A. Brebbia and J. Mackerle (Eds.). *Boundary Elements Communications*. ISBE, Computational Mechanics Publications, Southampton, U.K.
96. *BEASY*, Computational Mechanics, Ashurst Lodge, Ashurst, Southampton, England.

97. *BEST3D*, Prepared by Computational Mechanics Division, Department of Civil Engineering, State University at Buffalo and United Technologies, Pratt and Whitney, Engineering Division, Prepared for National Aeronautics and Space Administration, Lewis Research Center, Cleveland, Ohio.
98. *GPBEST*, Boundary Element Software Technology Corporation, Getzville, New York.
99. *SYSNOISE*, LMS International, Interleuvenlaan 68, 3001 Leuven, Belgium.
100. *BEMAP*, Spectronics, Inc., Lexington, Kentucky.
101. *COMET/BEA*, Automated Analysis Corporation, Ann Arbor, Michigan.
102. J. Vivoli and P. Filippi. Eigenfrequencies of thin places and layer potentials. *Journal of the Acoustical Society America* **97**, 1127–1248 (1974).
103. G. DeMay. Calculation of eigenvalues of the Helmholtz equation by an integral equation. *International Journal of Numerical Methods in Engineering* **10**, 59–66 (1976).
104. G. DeMay. A simplified integral equation method for the calculation of the eigenvalues of the Helmholtz equation. *International Journal of Numerical Methods in Engineering* **11**, 1340–1342 (1976).
105. J.R. Hutchinson. Determination of membrane vibrational characteristics by the boundary integral equation methods, In: C.A. Brebbia, (Ed.), *Recent Advances in Boundary Element Methods*. Pentech Press, London, England, pp. 301–316 (1978).
106. J.R. Hutchinson and G.K.K. Wong. The boundary element method for plate vibrations. *Proceedings of the ASCE 7th International Confernce on Electronic Computation*, St. Louis, Mo., pp. 297–311 (1979).
107. G.K.K. Wong and J.R. Hutchinson. An improved boundary element method for plate vibrations. *Proceedings of the International Seminar Boundary Element Methods.*, Irvine, California, July (1981).
108. G.R.C. Tai and R.P. Shaw. Helmholtz equation eigenvalues and eigenmodes for arbitrary domains. *Journal of the Acoustical Society of America* **56**, 796–804 (1979).
109. R.P. Shaw. Boundary integral equation methods applied to wave problems. In: P. K. Banerjee and R. Butterfield (Eds.), *Developments in Boundary Element Methods 1*. Elsevier Applied Science Publishers, London, England, pp. 121–153 (1979).
110. Y. Niwa, S. Kobayashi and M. Kitahara. Determination of eigenvalue by boundary element methods. In: P.K. Banerjee and R.P. Shaw (Eds.), *Developments in Boundary Element, Methods* **2**, Chapter 7. Applied Science Publishers, London, pp. 143–176 (1982).
111. J.R. Hutchinson. Boundary methods for time dependent problems. *Proceedings of the 5th Engineering Mechanics Divison Conference*, ASCE, University of Wyoming, pp. 136–139 (1984).
112. J.R. Hutchinson. An alternative BEM formulation applied to membrane vibration. *Proceedings of the 7th International Conference on BEM*, Lake Como, Italy, Springer-Verlag, Berlin, pp. 6-13–6-25 (1985).
113. J.O. Adeye, M.J.H. Bernal and K.E. Pitman. An improved boundary integral equation method for Helmholtz equation, *International Journal of Numerical Methods of Engineering* **21**, 779–787 (1985).
114. J. Zhou. Computations of eigenfunctions and eigenfrequencies of two-dimensional vibrating structures by the boundary element method. *Proceedings of the 20th Conference on Decision and Control*. IEEE, pp. 2045–2049 (1989).
115. G. Bezine. A mixed boundary integral-finite element approach to plate vibration problems. *Mechanical Research Communications* **7**, 141–150 (1980).
116. D. Nardini and C.A. Brebbia. A new approach to free vibration analysis using boundary elements. In: C.A. Brebbia (Ed.), *Proceedings of the 4th International Conference on BEM, Southampton, England*, Springer-Verlag, Berlin, pp. 313–326 (1982).
117. D. Nardini and C.A. Brebbia. A new approach to free vibration analysis using boundary elements, *Applied Mathematical Modelling* **1**, 157–162 (1983).
118. D. Nardini and C.A. Brebbia. Boundary integral formulation of mass matrices for dynamic analysis. In: C.A. Brebbia (Ed.), *Topics in Boundary Element Research* **2**, Springer, Berlin, 152–169 (1985).
119. D. Nardini and C.A. Brebbia. The solution of parabolic and hyperbolic problems using an alternative boundary element formulation. In: C.A. Brebbia and G. Maier (Eds.), *Proceedings of the 7th International Conference on BEM* **1**, Italy. Springer, Berlin, pp. 387–397 (1985).

120. D. Nardini and C.A. Brebbia. Transient boundary element elastodynamics using the dual reciprocity method and modal superposition. In: M. Tanaka and C.A. Brebbia (Eds.), *Boundary Elements VIII, Proceedings of the 8th International Conference I*. Computational Mechanics Publications, Southampton, and Springer-Verlag, Berlin, pp. 435–443 (1986).
121. P.W. Partridge and C.A. Brebbia. The dual reciprocity boundary element method for the Helmholtz equation. In: C.A. Brebbia and A. Chaudouet-Miranda (Eds.), *Proceedings of the International Boundary Elements*. Computational Mechanics Publishers/Springer-Verlag, pp. 543–555 (1990).
122. A. Kanarachos and Ch. Provatidis. Performance of mass matrices for the BEM dynamic analysis of wave propagation problems. *Computational Methods of Applied Mechanical Engineering* **63**, 155–165 (1987).
123. S. Ahmad and P.K. Banerjee. Free vibration analysis using BEM particular integrals. *Journal of Engineering Mechanical ASCE* **112**, 682–695 (1986).
124. P.K. Banerjee, S. Ahmad and H.C. Wang. A new BEM formulation for the acoustic eigenfrequency analysis. *International Journal of Numerical Methods of Engineering* **26**, 1299–1309 (1988).
125. H.C. Wang and P.K. Banerjee. Axisymmetric free-vibration problems by the boundary element method. *Journal of Applied Mechanical ASME* **55**, 437–442 (1988).
126. H.C. Wang and P.K. Banerjee. Free-vibration analysis of axisymmetric solids by BEM. *International Journal of Numerical Methods of Engineering* **29**, 985–1001 (1990).
127. R.B. Wilson, N.M. Miller and P.K. Banerjee. Free-vibration analysis of three-dimensional solids by BEM. *International Journal of Numerical Methods of Engineering* **29**, 1737–1757 (1990).
128. J.P. Agnantiaris, D. Polyzos and D.E. Beskos. Three-dimensional structural vibration analysis by the dual reciprocity BEM. *Computational Mechanics* **21**, 372–381 (1998).
129. J.P. Agnantiaris, D. Polyzos and D.E. Beskos. Free vibration analysis of non-axisymmetric and axisymmetric structures by the dual reciprocity BEM. *Engineering of Analysis Boundary Elements* **25**, 713–723 (2001).
130. M. Kögl and L. Gaul. Free vibration analysis of anisotropic solids with the boundary element method. *Engineering of Analysis Boundary Elements* **27**, 107–114 (2003).
131. A. Ali, C. Rajakumar and S.M. Yunus. On the formulation of the acoustic boundary element eigenvalue problems. *International Journal of Numerical Methods Engineering* **31**, 1271–1282 (1991).
132. C. Rajakumar, A. Ali and S.M. Yunus. Lanczos algorithm for acoustic boundary element eigenvalue problems. *Journal of Acoustic Society America* **91**(2), 939–948 (1992).
133. J.P. Coyette and K.R. Fyfe. An improved formulation for acoustic eigenmode extraction from boundary element models. *Journal of Vibration Acoustic, Transactions of the ASME* **112**, 392–398 (1990).
134. R.A. Bialecki, P. Jurgaś and G. Kuhn. Dual reciprocity BEM without matrix inversion for transient heat conduction. *Engineering of Analysis Boundary Elements* **26**, 227–236 (2002).
135. S.T. Raveendra and P.K. Banerjee. Polynomial particular solutions based boundary element analysis of acoustic eigenfrequency problems. *International Journal of Numerical Methods Engineering* **35**, 1787–1802 (1992).
136. C. Rajakumar and A. Ali. Acoustic boundary element eigenproblem with sound absorption and its solution using Lanczos algorithm. *International Journal of Numerical Methods Engineering* **36**, 3957–3972 (1993).
137. C. Rajakumar, A. Ali and S.M. Yunus. Boundary element-finite element coupled eigenanalysis of fluid–structure systems. *International Journal of Numerical Methods Engineering* **39**, 1625–1634 (1996).
138. A.J. Nowak. Temperature fields in domains with heat sources using boundary only formulations. In: C.A. Brebbia (Ed.), *Proceedings of the 10th BEM Conference* **2**, Springer-Verlag/Computational Mechanics Publishers., pp. 233–247 (1988).
139. A.J. Nowak and C.A. Brebbia. The multiple reciprocity method. A new approach for transforming BEM domain integrals to the boundary. *Engineering of Analysis Boundary Elements* **6**, 164–167 (1989).

140. A.J. Nowak and C.A. Brebbia. Solving Helmholtz equation by boundary elements using the multiple reciprocity method. *Computers and Experiments in Fluid Flow*, In: G.M. Carlomagno and C.A. Brebbia (Eds.), Computational Mechanics Publishers and Springer-Verlag, Berlin, pp. 165–179 (1989).
141. N. Kamiya and E. Andoh. Robust boundary element scheme for Helmholtz eigenvalue equation, In: C.A. Brebbia and G.S. Gipson (Eds.), *Proceedings of the BEM* **13**, Computational Mechanics Publishers and Elsevier Sci. Pub., London, pp. 839–850 (1991).
142. N. Kamiya and E. Andoh. Standard eigenvalue analysis by boundary-element method. *Communications in Numerical Methods in Engineering* **9**, 489–495 (1993).
143. N. Kamiya, E. Andoh and K. Nogae. Eigenvalue analysis by boundary element method: new developments. *Engineering of Analysis Boundary Elements* **12**, 151–162 (1993).
144. S.M. Kirkup and S. Amini. Solution of the Helmholtz eigenvalue problem via the boundary element method. *International Journal of Numerical Methods Engineering* **36**, 321–330 (1993).
145. D. Polyzos, Dassios and D.E. Beskos. On the equivalence of dual reciprocity and particular integral approaches in the BEM. *Bound. Elms. Comm.*, **5**, 285–288 (1994).
146. T.W. Davies and F.A. Moslehy. Modal analysis of plates using the dual reciprocity boundary element method. *Engineering of Analysis Boundary Elements* **14**, 357–362 (1994).
147. N. Kamiya, K. Nogae and E. Andoh. Adaptive boundary elements for eigenvalue analysis of the Helmholtz equation. *Engineering of Analysis Boundary Elements* **14**, 211–218 (1994).
148. G. Davì and A. Milazzo. A symmetric and positive definite variational BEM for 2-D free vibration analysis. *Engineering of Analysis Boundary Elements* **14**, 343–348 (1994).
149. J.R. Chang, R.F. Liu, S.R. Kuo and W. Yeih. Application of symmetric indirect Trefftz method to free vibration problems in 2D. *International Journal of Numerical Methods Engineering* **56**, 1175–1192 (2003).
150. S.M. Niku and R.A. Adey. Computational aspect of the dual reciprocity method for dynamics. *Engineering of Analysis Boundary Elements* **18**, 43–61 (1996).
151. J.T. Chen and F.C. Wong. Analytical derivations for one-dimensional eigenproblems using dual boundary element method and multiple reciprocity method. *Engineering of Analysis Boundary Elements* **20**, 25–33 (1997).
152. W. Yeih, J.T. Chen and C.M. Chang. Applications of dual MRM for determining the natural frequencies and natural modes of an Euler–Bernoulli beam using the singular value decomposition method. *Engineering of Analysis Boundary Elements* **23**, 339–360 (1999).
153. J.R. Chang, W. Yeih and J.T. Chen. Determination of the natural frequencies and natural modes of a rod using the dual BEM in conjunction with the domain partition technique. *Computational Mechanics* **24**, 29–40 (1999).
154. J.T. Chen, C.X. Huang and K.H. Chen. Determination of spurious eigenvalues and multiplicities of true eigenvalues using the real-part dual BEM. *Computational Mechanics* **24**, 41–51 (1999).
155. J.T. Chen. Recent development of dual BEM in acoustic problems. *Computer Methods in Applied Mechanics and Engineering* **188**, 833–845 (2000).
156. S.R. Kuo, J.T. Chen and C.X. Huang. Analytical study and numerical experiments for true and spurious eigensolutions of a circular cavity using the real-part dual BEM. *International Journal of Numerical Methods Engineering* **48**, 1401–1422 (2000).
157. J.T. Chen, I.L. Chung and I.L. Chen. Analytical study and numerical experiments for true and spurious eigensolutions of a circular cavity using an efficient mixed-part dual BEM. *Computational Mechanics* **27**, 75–87 (2001).
158. M.S. Ingber, A.A. Mammoli and M.J. Brown. A comparison of domain integral evaluation techniques for boundary element methods. *International Journal of Numerical Methods Engineering* **52**, 417–432 (2001).
159. G.S. Gipson. *Boundary Element Fundamentals — Basic Concepts and Recent Developments in the Poisson Equation*. Computational Mechanics Publications, Southampton, U.K. and Boston, U.S.A.; In: C.A. Brebbia and J.J. Connor (Eds.), *Topics in Engineering*, Vol. 2 (1987).
160. *ANSYS® User's Manual* IV, Revision 5.6, ANSYS, Inc., Canonsburg, PA (1999).

161. T.A. Cruise. *Mathematical Foundations of the Boundary-Integral Equation Method in Solid Mechanics*. Air Force Office of Scientific Research: Special Scientific Report AFOSR-TR-77-1002 (1977).
162. J.C. Lachat. *A Further Development of the Boundary Integral Technique for Elastostatics*. Ph.D. Thesis, Southampton University (1975).
163. M. Abramowitz and I.A. Stegun (Eds.). *Handbook of Mathematical Functions*. Dover, New York (1965).
164. L.C. Wrobel. *Potential and Viscous Flow Problems Using the Boundary Element Method*. Ph.D. Thesis, Southampton University (1981).
165. L.C. Wrobel and C.A. Brebbia. Axisymmetric potential problems. In C.A. Brebbia (Ed.), *New Developments in Boundary Element Methods*. Butterworths, London (1980). CML Publications, Southampton (1980).
166. C. Rajakumar. A comparison of acoustic finite element solutions to theory. *ANSYS News*, Fourth issue, Swanson Analysis Systems, Inc., Houston, PA 15342 (1988).
167. C. Rajakumar, A. Ali and S.M. Yunus. A new acoustic interface element for fluid–structure interaction problems. *International Journal of Numerical Methods Engineering* **33**, 369–386 (1992).
168. Y. Iwasaki, H. Kawabe and M. Bessho. The underwater sound scattering problem from the floating elastic shell. In: C.A. Brebbia et al. (Eds.), *Boundary Elements IX, Vol. 3, Fluid Flow and Potential Applications*. Springer-Verlag, London, pp. 53–64 (1987).
169. I.C. Mathews. Numerical techniques for three-dimensional steady-state fluid–structure interaction. *Journal of the Acoustical Society of America* **79**, 1317–1325 (1986).
170. D.T. Wilton. Acoustic radiation and scattering from elastic structures. *International Journal of Numerical Methods Engineering* **13**, 123–138 (1978).
171. J.S. Patel. Radiation and scattering from an arbitrary elastic structure using consistent fluid structure formulation. *Comp. Struct.* **9**, 287–291 (1978).
172. G.C. Everstine and F.M. Henderson. Coupled finite element/boundary element approach for fluid structure interaction. *Journal of the Acoustical Society of America* **87**, 1938–1947 (1990).
173. E.A. Schroeder and M.S. Marcus. *Finite Element Solution of Fluid–Structure Interaction Problems*. 46th Shock and Vibration Symposium, San Diego, CA, 1–18, Oct. (1975).
174. D. Young. Vibrations of rectangular plates by the Ritz method, *Journal of Applied Mechanics, Trans. ASME* **72**, 448–453 (1950).
175. S. Timoshenko and S. Woinowsky-Krieger. *Theory of Plates and Shells*, 2nd ed. McGraw-Hill, New York (1959).
176. T. Shuku and K. Ishihara. The analysis of the acoustic field in irregularly shaped room by the finite element method. *Journal of Sound and Vibration* **29**, 67–76 (1973).
177. D.J. Nefske, J.A. Wolf and L.J. Howell. Structural-acoustic finite element analysis of the automobile passenger compartment: a review of current practices. *Journal of Sound and Vibration.* **80**, 247–266 (1982).
178. C. Rajakumar and C. Rogers. The Lanczos algorithm applied to unsymmetric generalized eigenvalue problems. *International Journal of Numerical Methods Engineering* **32**, 1009–1026 (1991).
179. C. Rajakumar. Lanczos algorithm for the quadratic eigenvalue problem in engineering applications. *Computer Methods in Applied Mechanics and Engineering* **105**, 1–22 (1993).
180. M. Moller and S. Stewart. An algorithm for solving the generalized matrix eigenvalue problems. *Siam Journal of Numerical Analysis* **10**, 241–256 (1973).
181. A. Leissa and Z. Zhang. On the three-dimensional vibrations of the cantilevered rectangular parallelepiped. *Journal of the Acoustical Society of America* **73**, 2013–2021 (1983).
182. B.S. Garbow. EISPACK—For the real generalized eigenvalue problems. *Report*. Applied Mathematics Division, Argonne National Laboratory, U.S.A. (1980).
183. W.H. Yang. A method for eigenvalues of sparse-matrices. *International Journal of Numerical Methods Engineering* **19**, 943–948 (1983).
184. N. Kamiya, E. Ando and K. Nogae. A new complex-valued formulation and eigenvalue analysis of the Helmholtz equation by boundary element method. *Advances in Engineering Software* **26**, 219–227 (1996).

185. W. Yeih, J.T. Chen, K.H. Chen and F.C. Wong. A study on the multiple reciprocity method and complex-valued formulation for the Helmholtz equation. *Advances in Engineering Software* **29**, 7–12 (1998).
186. O.C. Zienkiewicz and R.E. Newton. Coupled vibrations of a structure submerged in compressible fluid. *Proceedings of International Symposium on Finite Element Techniques*. Stuttgart (1969).
187. L.E. Kinsler, A.R. Frey, A.B. Coppens and J.V. Sanders. *Fundamentals of Acoustics*. Wiley, New York, pp. 210–214 (1982).
188. I.-W. Yu. Subspace iteration for eigen-solution of fluid–structure interaction problems. *ASME Journal of Pressure Vessel Technology* **109**, 244–248 (1987).
189. *Mark's Standard Handbook for Mechanical Engineers*. McGraw-Hill, New York, 6-124, 6-163 (1978).
190. M.A. Jones, L.A. Binks and D.J. Henwood. Finite element methods applied to the analysis of high-fidelity loudspeaker transducers. *Computer and Structures* **44**, 765–772 (1992).
191. A. Craggs. A finite element model for acoustically lined small rooms, *Journal of Sound and Vibration* **108**(2), 327–337 (1986).
192. J.H. Wilkinson. *The Algebraic Eigenvalue Problem*. Clarendon Press, Oxford (1988).
193. J. Cullum and R.A. Willoughby. A practical procedure for computing eigenvalues of large sparse non-symmetric matrices. In: J. Cullum and R.A. Willoughby (Eds.), *Large Scale Eigenvalue Problems*, North-Holland, Amsterdam (1986).
194. Y. Saad. The Lanczos biorthogonalization algorithm and other oblique projection methods for solving large unsymmetric systems. *SIAM Journal of Numerical Analysis* **19**, 485–506 (1982).
195. B. Nour Omid. The lanczos algorithm for solution of large generalized eigenvalue problems. In: T.J.R. Hughes (Ed.), *The Finite Element Method*. Prentice-Hall, Englewood Cliffs, New Jersy (1987).
196. H.M. Kim and R.R. Craig, Jr., Structural dynamic analysis using an unsymmetric block Lanczos algorithm. *International Journal of Numerical Methods Engineering* **26**, 2305–2318 (1988).
197. *ANSYS® Engineering Analysis Systems User's Manual*, Revision 5.6. Ansys, Inc. Systems, Inc., Canonsburg, PA (1999).
198. M. Borri and P. Mantegazza. Efficient Solution of Quadratic Eigenproblems Arising in Dynamic Analysis of Structures. *Computer Methods in Applied Mechanics and Engineering* **12**, 19–31 (1977).
199. A. Ali and D. Ostergaard. Implementation of FE–BE hybrid techniques into finite element programs. In: S. Kobayashi and N. Nishimura (Eds.), Boundary Element Methods: Fundamentals and Applications. *Proceedings of IABEM Symposium*, Kyoto, Japan, October 14–17, 1991. Springer-Verlag, Berlin, pp. 11–20 (1992).
200. M.A. Golberg, C.S. Chen, H. Bowman and H. Power. Some comments on the use of radial basis functions in the dual reciprocity method. *Computational Mechanics* **21**, 141–148 (1998).
201. M.A. Golberg, C.S. Chen and H. Bowman. Some recent results and proposals for the use of radial basis functions in the BEM. *Engineering of Analysis Boundary Elements* **23**, 285–296 (1999).
202. M.D. Mikhailov. Integrals of radial basis functions for boundary element method. *Communications in Numerical Methods in Engineering* **16**, 683–685 (2000).

Index

For Product Safety Concerns and Information please contact our EU representative GPSR@taylorandfrancis.com Taylor & Francis Verlag GmbH, Kaufingerstraße 24, 80331 München, Germany

Batch number: 10725802

Printed by Printforce, the Netherlands